7th International Multi-Conference on Engineering and Technology Innovation (IMETI2018)

Recent Advances in Intelligent Technology Innovation

IOP Conference Series: Materials Science and Engineering
Volume 658

Taoyuan, Taiwan
2 - 6 November 2018

ISBN: 978-1-5108-9851-6
ISSN: 1757-8981

Printed from e-media with permission by:

Curran Associates, Inc.
57 Morehouse Lane
Red Hook, NY 12571

Some format issues inherent in the e-media version may also appear in this print version.

Institute of Physics
Dirac House, Temple Back
Bristol BS1 6BE UK

Phone: 44 1 17 929 7481
Fax: 44 1 17 920 0979

techtracking@iop.org

Additional copies of this publication are available from:

Curran Associates, Inc.
57 Morehouse Lane
Red Hook, NY 12571 USA
Phone: 845-758-0400
Fax: 845-758-2633
Email: curran@proceedings.com
Web: www.proceedings.com

Table of contents

Volume 658

International Multi-Conference on Engineering and Technology Innovation

2–6 November 2018, Taoyuan, Taiwan

Accepted papers received: 30 September 2019
Published online: 28 October 2019

Preface

Papers

IOP Conf. Series: Materials Science and Engineering **658** (2019) 011001 doi:10.1088/1757-899X/658/1/011001

The 7th International Multi-Conference on Engineering and Technology Innovation 2018 (IMETI2018) scheduled from November 02-06, 2018 in Taoyuan, Taiwan is an expansion of the International Conferences on Engineering and Technology Innovation (ICETI2011, ICETI2012, ICETI2014), and International Multi-Conference on Engineering and Technology Innovation (IMETI2015, IMETI2016, IMETI2017).

The conference is mainly organized by Taiwan Association of Engineering and Technology Innovation. The theme of the conference is set as "Recent Advances on Intelligent Technology Innovation". IMETI2018 consists of two sub-conferences (ICBEI2018 and ICECEI2018), 15 regular sessions, and 25 special sessions. This is a great opportunity for the engineering and technology community to present their new research results, exchange ideas and become familiar with new trends and directions in the fields of Science and Engineering Innovation. In addition, it is also a unique opportunity for us to form an international collaboration framework for promoting engineering and technology innovation.

We have with us today participants from research funding organizations, universities, as well as other research agencies. I hope that this five-day conference - a platform for thought-leaders, academics and researchers to share their ideas and views on common engineering and technology innovation issues - will challenge all participants, in particular the heads of institutions, to think more about the technology challenges and responses in engineering innovation which may in turn inspire new and practicable standards in the field.

The papers submitted from 16 countries on various topics are presented in various parallel oral and poster sessions in the conference. To all members of the Technical and Organizing Committees, we would like to take this opportunity to thank all of them for their tremendous efforts to organize this conference successfully.

The organizing committee of the conference thanks all the participants for their fruitful work and personal contribution to the development of these conference proceedings.

IOP Conf. Series: Materials Science and Engineering **658** (2019) 011002 doi:10.1088/1757-899X/658/1/011002

Committees

Program Chairman

- Wen-Hsiang Hsieh, President, Taiwan Association of Engineering and Technology Innovation
  National Formosa University, Taiwan
- Zongyu Chang, Ocean University of China, China

Program Co-Chairman

- Chun-Pin Lin, Asia University, Taiwan
- Ing-Jr Ding, National Formosa University, Taiwan
- Jia-Shing Sheu, National Taipei University of Education, Taiwan
- Jian-Ming Lu, National Center for High-Performance Computing, National Applied Research Laboratories, Taiwan
- Van-Tsai Liu, National Formosa University, Taiwan

Technical Committee Chairman

- Chih-Ta Yen, National Formosa University, Taiwan
- Tao-Hsing Chen, National Kaohsiung University of Applied Sciences, Taiwan
- Yi-Chang Wu, National Yunlin University of Science & Technology, Taiwan

Program Committee

- An Cheng, National Ilan University, Taiwan
- Chan Young Park, Hallym University, South Korea
- Chang-Chun Lee, Chung Yuan Christian University, Taiwan
- Chang-Huang Chen, Tungnan University, Taiwan
- Chen Lu, Beihang University, China
- Chun-Hsiung Tseng, Nanhua University, Taiwan
- Chung-Chi Huang, Far East University, Taiwan
- David Pei-Cheng Lin, Chung Shan Medical University, Taiwan
- Dae-Jin Kim, Kyung Hee University, South Korea
- Hsiang-Chen Hsu, I-Shou Univerity, Taiwan
- Hsiao-Han Liu, I-Shou University, Taiwan
- Hsin-Hung Lee, Center for Measurement Standards, Industrial Technology Research Institute, Taiwan
- Kuan-Chen Lin, Far East University, Taiwan
- Jeong-Dong Kim, Korea University, South Korea
- Jie-Tong Zou, National Formosa University, Taiwan
- Jih-Gau Juang, National Taiwan Ocean University, Taiwan
- Jium-Ming Lin, Chung Hwa University, Taiwan
- Minvydas Ragulskis, Kaunas University of Technology, Lithuania
- Shih-Sung Lin, National Defense University, Taiwan
- Tai-Ping Chang, National Kaohsiung first University of Science and Technology, Taiwan
- Te-Kwei Wang, Ming Chi University of Technology, Taiwan
- Wei-Ting Lin, National Ilan University, Taiwan
- Yi-Cheng Huang, National Changhua University of Education, Taiwan
- Yih-Chuan Lin, National Formosa University, Taiwan
- Yit-Jin Chen, Chung Yuan Christian University, Taiwan
- Yu Guo, Kunming University of Science and Technology, China
- Yung-Cheng Lee, WuFeng University, Taiwan
- Yung-Hui Chen, Lunghwa University, Taiwan

International Scientific Committee

- A. Gottscheber, University of Heidelberg, Germany
- Amir Abidov, Tashkent State Technical University, Uzbekistan
- Aylie Han, University of Diponegoro, Indonesia
- A. Walton, University of Edinburgh, UK
- Buntara S. Gan, Nihon University, Japan
- R. S. Beniwal, National Institute of Science Communication and Information Resources, India
- C. Warwick, University College London, UK
- E. Manske, TU Ilmenau, Germany
- G. Jäger, TU Ilmenau, Germany
- Giovanni Pau, Kore University of Enna, Italy
- Hyeungsik Choi, Korea Maritime and Ocean University, Korea
- Igor Meglinski, University of Oulu, Finland
- J. Wood, Goldsmiths College, University of London,UK

- J. Carlos, Metrolho Instituto Politecnico de Castelo Branco Castelo Branco, Portugal
- Jong-Han Lee, Daegu University, South Korea
- J.S. Dai, Kings of college London, UK
- Kihong Shin, Andong National University, South Korea
- M. Woolley, University of Coventry, UK
- P.F. Savanna, College of Art and Design, USA
- R.X. Du, The Chinese University of Hong Kong, HK
- S.G. Hanson, Technical University of Denmark, Denmark
- S.D. Prior, University of Middle Sex, UK
- T. Froelich, TU Ilmenau, Germany
- Uldis Strautins, University of Latvia, Latvia
- WooSeok Kim, Chungnam National University, South Korea
- Zhengchao Xie, University of Macau, Macau

Advisory Committee

- B.W. Huang, Cheng Shiu University, Taiwan
- C.C. Wen, Metal Industries Research & Development Centre, Taiwan
- C.H. Huang, Wufeng University, Taiwan
- Chun-Huy Wang, Nan-Jeon University of Technology, Taiwan
- C.M. Lin, National Taitung University, Taiwan
- C.T. Lin, Metal Industries Research & Development Centre, Taiwan
- F. Long, Xiamen University, China
- H.S. Lin, National Kaohsiung University of Applied Sciences, Taiwan
- I.T. Tang , National University of Tainan, Taiwan
- J.C. Zhong, Changchun University of Science and Technology, China
- J.F. Yao, Xiamen University, China
- J.X. Xie University of Science and Technology, Beijing, China
- K.H. Wen, Fuzhou University, China
- K.S. Shu, Chia Nan University of Pharmacy and Science, Taiwan
- L.W. Ji, National Formosa University, Taiwan
- M. Chang, Chung Yuan Christian University, Taiwan
- M.H. Wang, National Chin-Yi University of Technology, Taiwan
- Q.K. Xue, Tsinghua University, Beijing, China
- S.J. Chang, National Cheng Kung University, Taiwan
- S.J. Lin, National Kaohsiung University of Applied Sciences, Taiwan
- S.T. Shen, National Formosa University, Taiwan
- Sungjin Kim, Kumoh National Institute of Technology, South Korea
- T.H. Meen, National Formosa University, Taiwan
- W. Water, National Formosa University, Taiwan
- W.Z. Guo, Shanghai Chia-Tung University, China
- Wonho Suh, Hanyang University, South Korea
- X.L. Du, Beijing National Laboratory for Condensed Matter Physics, China
- Y.C. Wang, National Yunlin University of Science & Technology, Taiwan
- Y.L. Hwang, National Formosa University, Taiwan
- Y.L. Yeh, Far East University, Taiwan
- Y.Q. Zhou, China University of Petroleum, China
- Y.X. Sun, China University of Petroleum, China

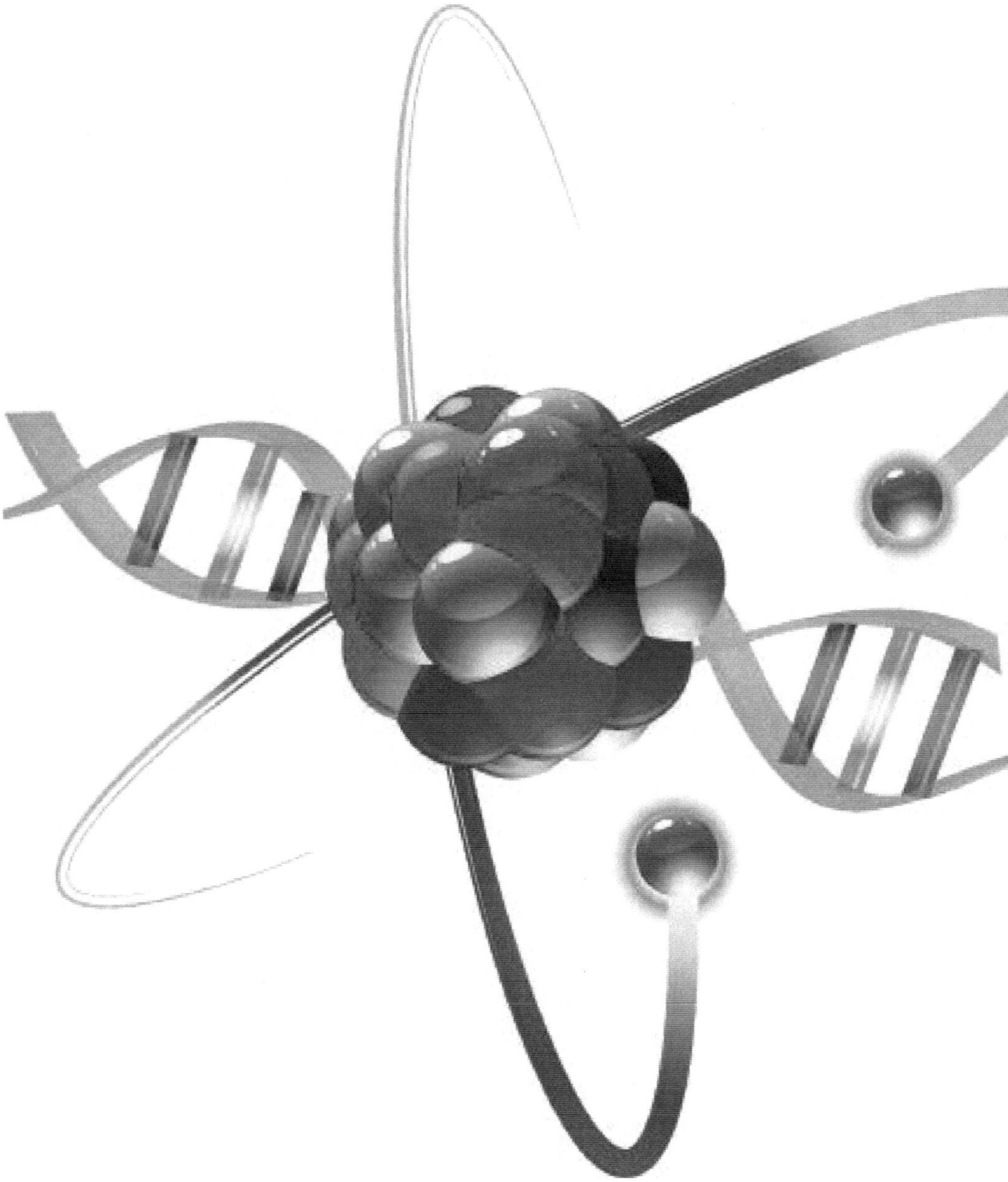

IOP Conf. Series: Materials Science and Engineering **658** (2019) 011004 doi:10.1088/1757-899X/658/1/011004

Peer review statement

All papers published in this volume of *IOP Conference Series: Materials Science and Engineering* have been peer reviewed through processes administered by the proceedings Editors. Reviews were conducted by expert referees to the professional and scientific standards expected of a proceedings journal published by IOP Publishing.

IOP Conf. Series: Materials Science and Engineering **658** (2019) 012001 doi:10.1088/1757-899X/658/1/012001

Effects of sequential pretreatment of rice straw and coconut shell for improved biomethane production

A Magomnang

Mechanical Engineering Department, University of Science and Technology of Southern Philippines – CDO Campus, Cagayan de Oro City, Philippines

E-mail: a_magomnang@yahoo.com

Abstract. This study aims to determine the effects of sequential pretreatment of selected agricultural biomass wastes in terms of biomass properties as well as its biomethane yield on the anaerobic co-digestion of wastes co-digested with various biomass feedstocks for power generation applications. This was carried out by determining the effects on the biomethane production using the two pre-treatments, that is 3% and 4% w/v of sodium hydroxide subject to ultrasonication and liquid hot water hydrolysis pretreatment for the rice straw and coconut shell feedstocks. As it will be used as substrates for the anaerobic co-digestion experiments; using digested manures as an inoculant. Results show that biomethane production increased by 140% and 290% from the pre-treatment of rice straw and coconut shells, respectively. With these, the pre-treated coconut shell subjected to Ultrasonication with 3% sodium hydroxide and liquid hot water shows the best effect among the pre-treatment of biomass feedstocks of rice straw and coconut shell at various sodium concentrations. The results of this experiment would give a viable estimate of the possible biomethane production from these agricultural wastes.

1. Introduction

At present, feed stocks from the municipal solid waste, industry, food processing plants and wastewater treatment plant, are limited; thus, new renewable sources are sought after. With the abundant availability of agricultural biomass in the Philippines makes it a highly interesting source of feedstock, from a variety of crops like palay, corn, sugarcane, cassava, and coconut, and from livestock manure of carabao, cattle, hog, goat and dairy, that can be utilized for biogas production. These sources can serve as fuel for power generation or as feedstock for advanced biofuels production. Given these abundant supplies of biomass in the country, it would be advantageous for the Philippines to exploit these resources to address the country's energy dependence, mitigate climate change and eventually achieve economic growth and prosperity.

Biogas is a renewable, high-quality fuel which can be produced from various organic raw materials and used for various energy services. Biogas technology has been developed and widely used over the world because it has several advantages – reduction of the dependence on non-renewable resources, high energy-efficiency, environmental benefits, available and cheap resources to feedstock, relatively easy and cheap technology for production, and extra values of digestate as a fertilizer [1]. However, the current status of biogas production and utilization varies largely among countries. To overcome the issues related to the utilization of lignocellulosic materials like rice straw and coconut shell for biogas production, new pretreatment strategies have been evaluated and developed. One of these strategies is the combination of sodium hydroxide hydrolysis, ultrasonication, and liquid hot water pretreatment.

Typically, the ultrasonication and liquid hot water pretreatment were used to modify the lignin and cellulose structure of the biomass to generate ethanol [2] but for this case, said pretreatments were done to improve biogas yield. In a study by sodium hydroxide hydrolysis was found to substantially increase the biogas yield of rice straw.

Currently, there are many researches on the pretreatment of biomass wastes similar to paddy straw using sodium hydroxide-microwave pretreatment [3], it was reported that 4% NaOH-30 minutes of microwave was found to be the best pretreatment which resulted in 65.0% decrease in lignin content and 88.7% reduction in silica content that resulted in 54.7% increase in biogas production [3]. Moreover, a mechanical disruption of algae using ultrasound homogenizer and thermo-alkaline pretreatment enzymatic hydrolysis demonstrated the best biogas yield than other pre-treatment which reached 626.5 mL/g COD with 62.65% of biodegradability [4]. Furthermore, liquid hot water pretreatment of giant reed in a Parr reactor and alkaline pretreatment was studied using the various NaOH concentrations which improve glucose yield that is significantly increased the cumulative methane yield by 63% than of the untreated biomass [5]. In addition, the ultrasonication of Cotton Gin Trash (CGT) followed by a liquid hot water pretreatment and lignolytic enzymes pretreatment combination generated high amounts of sugar that increased delignification and modified the cellulose structure of the CGT as confirmed by the FT-IR Spectrum [2].

However, a combination of these pretreatments (ultra-sonication, liquid hot water, and NaOH hydrolysis), has not been tested in rice straw (RS) and coconut shell (CS) biomass wastes. Thus, the synergic effect of these technologies may raise de-lignification, cellulose conversion, and biogas yield from RS & CS. The effect of these pretreatments over the RS & CS structure can be determined using the Fourier transform infrared (FT-IR) spectrum. In principle, FT-IR analysis is applied to study qualitatively the modifications in structure and is not utilized for quantitative analysis. However, it can be used to identify the modification in biomass structure after different pretreatments than the conventional structural examination. From a broad perspective, the requirements of the pretreatment are to (1) improve the accessibility of the enzymes to the cellulose and hemicelluloses and thereby the degradability, (2) avoid degradation or loss of the carbohydrates, (3) avoid the production of potential process inhibitors, (4) be cost and energy efficient, and (5) have low negative impact as much as possible on the environment [6].

Anaerobic co-digestion experiments were performed in this study to provide information on the long-term effects, such as eventual inhibition during the digestion of a lignocellulosic substrate. A co-digestion is often beneficial, since it supplies the system with more nutrients, leading to a better balance in the C/N ratio and in the pH. It also improves the stability of the process and increases methane yield due to positive synergistic effects thereby increasing the economic value of the biogas. This study aims to determine the effects of various pretreatment methods (alkalinity using NaOH, ultrasonication, liquid hot water using autoclave) on the biomass properties and biogas yield in the anaerobic co-digestion of agricultural biomass waste utilizing cattle (cow) manure co-digested with different biomass feedstocks such as rice straw, coconut shell, and sewage sludge as inoculum for power generation applications.

2. Materials and methods

2.1. Experimental procedure and analysis

In this study, effects of sequential alkaline-ultrasonication and hot water treatment on biogas production from rice straw and coconut shell were determined. The ultrasonication step was modified. It was simultaneously performed with basic hydrolysis using different concentrations (3 and 4% w/v) of sodium hydroxide (NaOH) solution. The ultrasonicator (Hielscher Ultrasonic Processors, Ringwood, NJ, USA) was set at the highest value of amplitude (100%) and cycle (1). The biomass was not washed prior to hot water treatment using the autoclave, thus remaining particles of NaOH can still be present adhering with the biomass. The liquid hot water pretreatment was conducted at 121°C, 1.02 atm for 1 hr. in an autoclave using a feedstock concentration of 10% solids. The experimental procedure and analysis conducted in this study are shown in Table 1. Control set-ups were conducted using unpretreated rice straw or coconut shell as a substrate in the digester. All the experiments were done in duplicates, with biogas yield and methane concentration as response variables.

Table 1. Summary of test procedures conducted in this study.

Test Procedures	ASTM Standards
Moisture Content	E 871 - 82 (2013)
Elemental Analysis	
Carbon and Hydrogen	E 777 - 08
Total Sulfur	E 775 - 15
Nitrogen	E 778 - 15
Oxygen	By Difference
Proximate Analysis	
VCM	E 872 - 82 (2013)
Ash	E1755 - 01(2015)
Fixed Carbon	By Difference
Heating Value	E 771

2.2. Pretreatment Experiments

The experiment followed the sequence of pretreatments such as ultrasonication [2, 4, 7], with alkaline treatment through NaOH simultaneously [3-5, 7]; and liquid hot water [2-3, 5]. As shown in Table 2, rice straw and coconut shell samples were subjected to mechanical disruption using the ultrasonication process. A solution containing 10% w/v solid biomass was put in an ultrasonicator (Hielscher Ultrasonic Processors, Ringwood, NJ, USA) with operating conditions set at the highest value of amplitude (100%) and cycle (1). Inside the reactor, ultrasonic waves create cavitation in the liquid medium (NaOH) which contains 250 grams of rice straw sample. Samples in 3,000 ml Erlenmeyer flask were manually stirred, then sonicated for 1 hr. After the ultrasonication process, the biomass was not washed prior to hot water treatment. The liquid hot water pretreatment was performed in an autoclave for 1 hr at a temperature of 121°C and working pressure of 15 psi. The slurry was drained off. The remaining solids were washed with deionized water until the washings were clean, colorless and the pH was neutral. The rice straw was then dried overnight in an oven at 105°C. Dried rice straw and coconut shell were stored in polyethylene bags and were used for proximate analysis (total solids, volatile solids, and ash). Untreated rice straw and coconut shell were also subjected to analysis to determine the extent of their degradation.

Table 2. Pretreatment conditions evaluated in this study.

Parameters	Rice Straw		Coconut Shell	
Weight of Biomass (grams)	250	250	250	250
NaOH Concentration (% w/v)	3	4	3	4
Ultrasonication Treatment Conditions	1 hr., 100%, cycle (1)	1 hr., 100%, cycle (1)	1 hr., 100%, cycle (1)	1 hr., 100%, cycle (1)
Liquid Hot Water Treatment Conditions	121°C, 1.02 atm, 1 hr.	121°C, 1.02 atm, 1 hr.	121°C, 1.02 atm, 1 hr.	121°C, 1.02 atm, 1 hr.

2.3. Anaerobic co-digestion experiments

Batch trials were performed to investigate the effects of biogas production from rice straw and coconut shell substrates subjected to pretreatments such as ultrasonication with NaOH simultaneously; and Liquid Hot water treatment. For each batch trial, reactors were filled with the inoculum, which is cattle manure, and sewage sludge; and co-digested simultaneously with various combinations of biomass like rice straw and coconut shell. Control reactors containing only cattle manure, and cattle manure with sewage sludge were maintained during each batch trial. Feed compositions used for the batch trials are given in Table 2. Each reactor had a total working volume of approximately 4.8 L, and duplicate reactors were used for each feed composition. The trials were performed in an environmental chamber maintained at 35°C. The digesters were mixed daily by turning them upside-down and shaking for about 20 seconds.

Biogas volumes were recorded daily. The gas collectors were marked to provide a direct reading of volume. Biogas was discharged from the gas collectors by lifting the overflow carboy above the collectors to refill the collectors with water and opening the valves on top of the digesters to allow the gas to exit. Gas sampling was done through the use of a tee connection located between the gas collector and digester which were sealed with a rubber stopper. The stopper was removed and a syringe was inserted into the tee to collect a gas sample. While gas was flowing out of the gas collector, the syringe was gradually opened to withdraw a sample. Approximately 60 mL was collected for each biogas sample which was analyzed immediately using gas chromatography unit (SRI Gas Chromatograph). Gas samples were collected daily during the batch trials. Liquid samples were taken only the first and last days of each batch trial to determine solids content characterization. The liquid samples were collected in duplicate, with approximately 75 mL of material collected for each sample. The liquid samples are subjected to moisture content determination, proximate, ultimate and heating value analysis.

3. Results and discussion

3.1. Biomass pretreatment

The rice straw and coconut shell were subjected to a sequence of pretreatments such as ultrasonication with alkaline treatment through NaOH simultaneously; liquid hot water through the autoclave. However, the ultrasonication step for this experiment was modified to simultaneously perform a basic hydrolysis using different concentrations of NaOH (3 and 4% w/v) and a control of untreated rice straw as shown in Table 1. These values were selected to avoid producing a waste chemical solution and also to reduce the pretreatment cost [8]. Another study in the literature reported that 3-4% is the optimal dose for five NaOH samples (2, 4, 6, 8, and 10%) when was used to pretreat rice straw digestibility and biogas production [3, 12]. Series of test and analysis for untreated and pretreated rice straw and coconut shells were conducted, such as % total solids and moisture content determination through proximate and ultimate analysis as shown in Table 3. The analysis shows VCM content in rice straw is significantly higher than the untreated ones. In fact, the untreated rice straw increased VCM content from 66.56% to 84.73% and 88.18%, for the 3% NaOH+U+LHW RS and 4% NaOH+U+LHW RS respectively. Also, the carbon content of the pretreated coconut shell is higher than the pretreated rice straw by 7.25 percentage points and 8.84 percentage points for 3% and 4% NaOH+U+LHW pretreatment respectively. The total solid content determination was conducted in duplicates, while moisture content determination was done in triplicates based on the ASTM E1756 standard.

As shown in Table 3, the pretreatment (using alkaline-ultrasonication and Liquid Hot water using autoclave) of rice straw and coconut shell with 3% w/v NaOH concentration (3% NaOH+U+LHW RS and 3% NaOH+U+LHW CS), yielded the highest heating value, followed by the 4% NaOH (4% NaOH+U+LHW RS and 3% NaOH+U+LHW CS). An additional examination was conducted on the ultimate, proximate analysis and heating value for the untreated and pretreated rice straw and coconut as well as the sewage sludge and cattle manure. Therefore, the said biomass samples are quite combustible in nature, as compared with the untreated rice straw. The amount of the Volatile Combustible Matter increases 30% more than the untreated ones. The said pretreatments can be interpreted that the lignin component of the rice straw is degraded during the pretreatments.

3.2. Biomass pretreatment

The effects of applying pretreatments using the Alkaline-Ultrasonication hydrolysis and Liquid Hot Water on the co-digestion of dairy cattle manure with rice straw and coconut shell, compared with the untreated residues, were determined. The said experiment undergoes 45 days of the anaerobic co-digestion process. The biogas production from rice straw and coconut shell subjected to four (4) pretreatments (i.e. Alkaline-Ultrasonication Hydrolysis and Liquid Hot water using autoclave pretreatments with varying NaOH concentration of 3% and 4% w/v) were monitored.

Table 3. Pretreatment conditions evaluated in this study.

Property	Rice Straw		Coconut Shell	
	3% w/v	4% w/v	3% w/v	4% w/v
VCM (%)	84.73	88.18	78.85	78.19
Ash (%)	11.26	4.79	18.78	19.55
Fixed Carbon (%)	4.01	7.03	2.37	2.26
C (%TS)	39.33	37.23	46.58	46.07
H (%TS)	5.76	5.48	5.77	5.64
O (%TS)	54.58	57.06	47.42	48.06
N (%TS)	0.04	0.02	0.02	0.02
S (%TS)	0.28	0.22	0.20	0.22
C:N Ratio	936.50	2326.56	1940.79	2879.19
Heating Value (MJ/kg)	15.76	13.08	19.28	19.33

Figure 1 shows that the pretreated rice straw produces higher biogas production than other pretreatments and the same effect was achieved in the pretreatment of the coconut shell, whereas 23.12 L of gas was produced (Table 4). The biogas production with co-digestion of rice straw is 1.89 times higher than the untreated ones and 1.50 times higher than the untreated coconut shell.

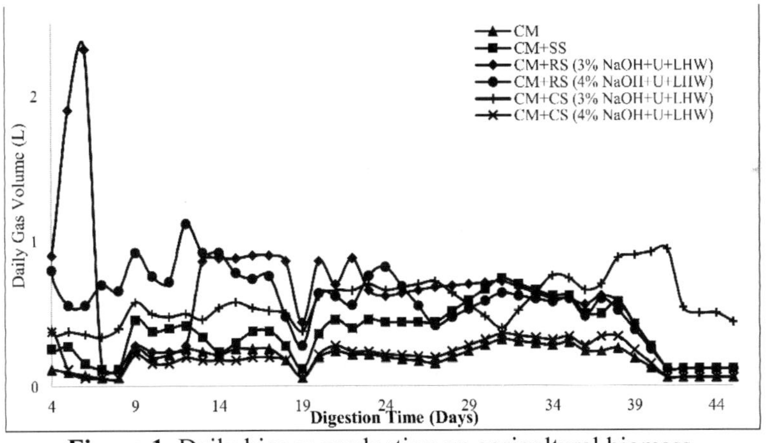

Figure 1. Daily biogas production on agricultural biomass.

Table 4. 45-days digestion results of pre-treated agricultural residues with cow manure.

Parameter	Cattle Manure & Rice Straw		Cattle Manure & Coconut Shell	
	3% w/v	4% w/v	3% w/v	4% w/v
C/N ratio	27.05	27.03	24.79	24.79
Biogas Yield (L)	18.18	12.54	23.12	8.44
Methane Yield (L)	9.88	6.96	13.84	4.77
Average Methane content (%)	53.80	53.00	55.90	54.50

Similarly, the percent methane concentration on the different treatments shown in Figure 2 maintains in the 50% mark except for the cow manure which slightly decreased as observed on 10th to 12th day. Likewise, similar reports were reported by [9] in the anaerobic co-digestion of cow manure at various maceration times which resulted in a methane concentration of 58% from the co-digestion of cow manure with rice straw. Moreover, the average percent methane concentration for these treatments presented in Figure 2 are also in the 50% mark as compared to the untreated rice straw and coconut shell that it took 25 days in order to reach the 50% methane concentration. Likewise, the methane concentration on all the biogas treatments reaches above 50% on its fourth day and readings maintain with slight variations and reported same levels with other publications [9-10]. This implies that fermentation/anaerobic digestion process properly converts the waste organic matter into a mixture of carbon dioxide and methane gas [3]. Methane production trends of the treated and pretreated biomass shown in Figure 3 follows similar production trends for all the digesters. In 5 days, methane production rates using treated biomass had increased over the initial production rates (Day 4) compared with the untreated ones for more than 300%. Hence, co-digestion of pretreated biomass helps to overcome the deficiencies of mono-digestion, which typically become the rate limiting step for the AD process [11-12]. Thus, anaerobic co-digestion of different organic materials may enhance the stability of the anaerobic processes [13] because of better carbon-to-nitrogen balance. In addition, co-digestion of rice straw and coconut shell at 4% w/v reduce the methane yield as shown in table 4 due to the inhibitory effect of high ammonia and sulfide concentrations [13] and exhibits a more stable biogas production.

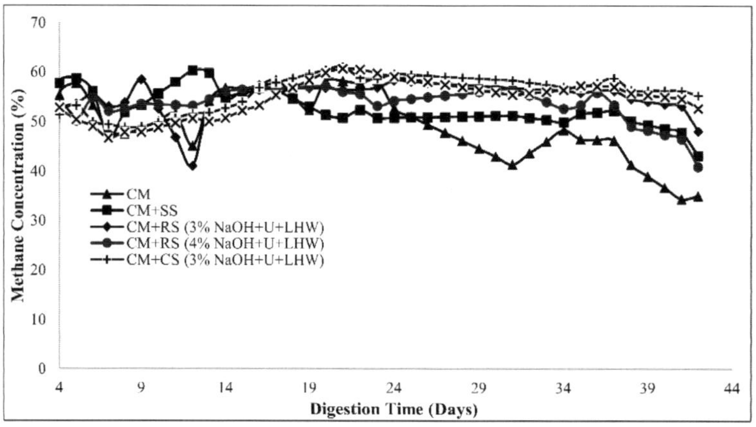

Figure 2. Methane concentration after 45-day anaerobic digestion of cattle manure, treated rice straw & coconut shell.

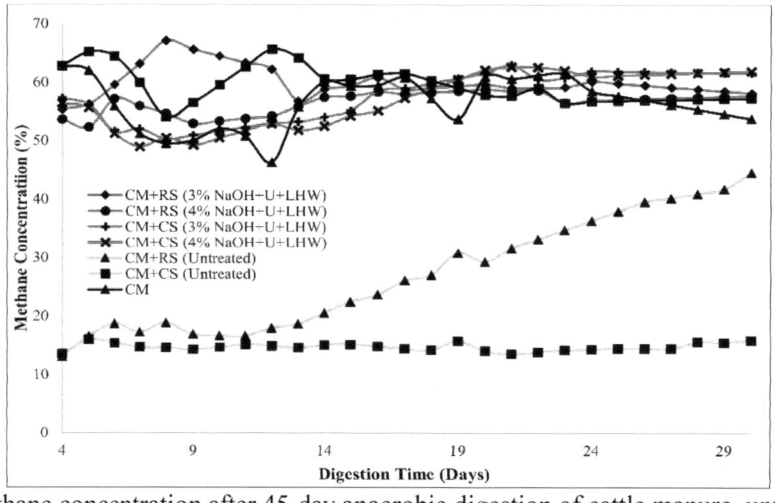

Figure 3. Methane concentration after 45-day anaerobic digestion of cattle manure, untreated/treated Rice Straw (RS) & Coconut Shell (CS).

4. Conclusion

This study focuses on the possible means of increasing methane production potential from anaerobic co-digestion of agricultural biomass waste. In agreement with results presented, it showed a positive impact on biogas production rate by biomass pretreatment using NaOH-Ultrasonication and Liquid Hot water pretreatment. The alkali-ultrasonication-liquid hot water pretreatment improved the conversion and resulted in methane yields increase of 140% & 290% from anaerobic co-digestion of rice straw and coconut shell with cow manure and sewage sludge as inoculum compared with untreated material. This study also shows how substrate composition, pretreatment methods, and operational parameters during anaerobic digestion affect the microbial consortia working in the digester.

References

[1] Bagher A M 2015 *Bull of Ad Sci Res*. **1**(5) pp 132 – 135
[2] Plácido J, Imam T and Capareda S 2013 *Bioresour technol* **139** pp 203 – 208
[3] Kaur K and Phutela U G 2016 *Ren Energ* **92** pp 178 – 84
[4] Karray R, Hamza M and Sayadi S 2015 *Bioresour technol* **187** pp 205 – 213
[5] Jiang D, Ge X, Zhang Q and Li Y 2016 *Bioresour technol* **216** pp 60 – 68
[6] Dehghani M, Karimi K and Sadeghi M 2015 *Energ & Fuel* **29**(6) pp 3770 – 3775
[7] Oz N A and Uzun A C 2015 *Ultrason sonochem* **22** pp565 – 572
[8] Zhang Y, Chen X, Gu Y and Zhou X 2015 *Appl Energ* **15** pp 39 – 48
[9] Silvestre G, Gómez M P, Pascual A and Ruiz B 2013 *Wat Sci Tech* **67**(4) pp 745 – 755
[10] Gu Y, Chen X, Liu Z, Zhou X and Zhang Y 2014 *Bioresour technol* **158** pp149 – 55
[11] Zhang W et al 2014 *Appl Energ* **128** pp 175 – 183
[12] Lianhua L et al 2010 *Inter Jour hydro Energ* **35**(13) pp 7261 – 7266
[13] Zhang C, Xiao G, Peng L, Su H and Tan T 2013 *Bioresour technol* **129** pp 170 – 176

Innovative and rapid design for I2C (inter-integrated circuit) interface communication of FPGA chips

Cheng Shion Shieh

Department of Electronic Engineering, Far East University, Taiwan

E-mail: csshieh@mail.feu.edu.tw

Abstract. A new design flow is proposed to greatly improve the efficiency and accuracy of the HDL program design. This paper successfully completes the FPGA (Field-Programmable Gate Array) chip's external communication I2C interface design of new design flow not traditionally one. FPGA chip is often used in mobile phones, military, aerospace or other industries. Even INTEL and Microsoft and other large companies have begun to acquire FPGA-related companies, showing its importance. But the FPGA language is written in the Hardware Description Language (HDL) which is a mechanical engineering staff obstacles. Therefore, this article makes use of the MATLAB / SIMULINK tool, which is familiar with the engineering staff, to complete the FPGA chip's I2C interface design. The advantage is that the functional simulation and HDL program can be completed at the same time, do not have to perform the function after the simulation before designing HDL program. The traditional design wastes many times to revise back and forth between function simulation and HDL program design. Finally an experimental illustration is proposed to verify the proposed design of the I2C communication interface for an HDL program on the read and write experiments of Electrically-Erasable Programmable Read-Only Memory (EEPROM).

1. Introduction

With the rapid development of electronic technology, many ICs need to communicate with each other. For this reason, in 1982 Philips Semiconductors of the Netherlands developed a bidirectional I2C (Inter-Integrated Circuit) bus [1] to reduce the number of microcontrollers or CPUs to connect many low-speed peripherals with fewer pins. Its advantages are simplicity and effectiveness. The purpose of the initial development of the I2C bus was to develop audio and video equipment that could be more easily connected to the CPU. Today, it has become the standard for inter-chip low-speed serial communication and is widely used in entertainment, consumer electronics and control electronics. [2].

Recently, FPGA chips have been widely used in industrial control [3-5], mobile phones, artificial intelligence, deep learning [6]. It shows the importance of FPGA chip design that Intel and Microsoft's technology companies have made great efforts to acquire related FPGA companies. FPGA chips often need to communicate with peripheral I2C-interfaced chips. Although there are specialized I2C bus interface chips on the market, most of them have fixed performance, single function, and inconvenient use. Therefore, it is necessary to redesign an HDL program that can implement the I2C bus interface read/write function on the FPGA chip according to the I2C bus protocol. HDL programming is an obstacle for most engineers because it is time-consuming and error-prone. Not easy to complete.

Most of the engineers are more familiar with the graphic control software Matlab/Simulink [7], which can directly simulate the program design right or wrong. As long as the simulation of the

IOP Conf. Series: Materials Science and Engineering **658** (2019) 012002 doi:10.1088/1757-899X/658/1/012002

Matlab/Simulink program is verified, this Matlab/Simulink program can be directly converted to the corresponding HDL code. Then use FPGA-specific software Quartus II to compile and download the HDL program. The design flow mentioned in this paper combines design and simulation to quickly achieve the desired control results which not only saves time but also helps to clarify of programming concepts.

This article is divided into: Section II describes the I2C communication bus agreement and design, the third section proposes the I2C's implementation by Simulink tool and QuartusII one, and the fourth section describes the I2C read design concept and proposes the implementation in Matlab/Simulink platform. Section 5 verifies FPGA's I2C Read/Write function on real circuit equipment.

2. I2C communication agreement and design [8]

I2C data communication consists of a data line (SDIN) and a clock line (SCLK). A master chip controls multiple slave ones. The master chip can be an FPGA, DSP, or MCU. The slave can be an audio codec chip, a TV Decoder chip, or an EEPROM chip. The schematic diagram is shown in Figure 1. The start condition of the I2C bus is that when both SCLK and SDIN are high, the SDIN first changes from high to low, and then SCLK changes from high to low. The stop condition is when both SCLK and SDIN signals are low, SCLK changes from low to high, and then SDIN changes from low to high, as shown in Figure 2.

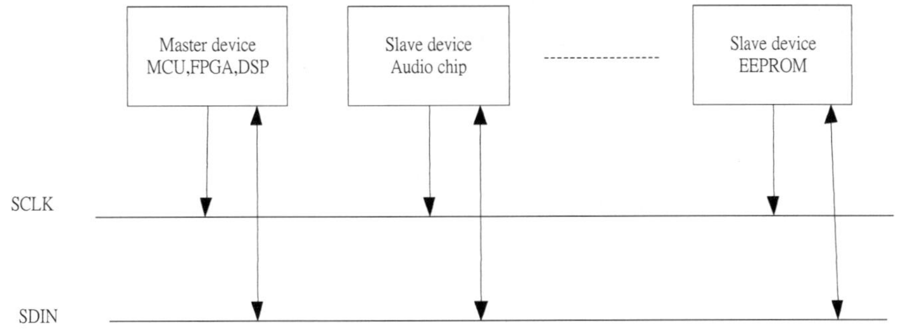

Figure 1. I2C master-slave device block.

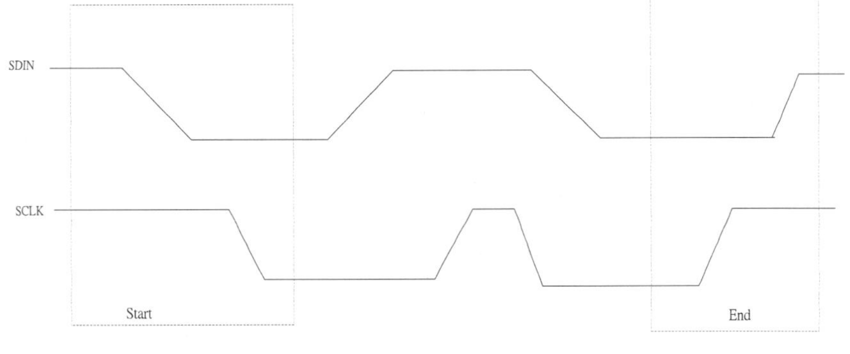

Figure 2. Start (ST) and End (SP) signals of the I2C bus.

The purpose of this paper is to develop a new I2C design process for master device FPGA. Verify the proposed process by firstly writing a piece of 8-bit data from the master device to the slave one, and then the master device FPGA will read the data from the slave device EEPROM to check master writing action right or wrong. That is, complete the write and read actions from the master device FPGA to the slave one EEPROM. This paper does not use the traditional program-writing method, but uses a Matlab/Simulink design in the form of graphic control design, and converts these graphic control programs into HDL codes, which are described in the following sections.

IOP Conf. Series: Materials Science and Engineering **658** (2019) 012002 doi:10.1088/1757-899X/658/1/012002

3. I2C write action of the master device

The master writes an 8-bit data to a certain register of the slave device. Firstly, find out the device address of the slave device and the address of the slave device register to be placed. The control style is that the master device generates a start signal ST. Then, the master control terminal sends a set of device address 7 bits to be controlled, plus a write (W) of 1 bit (write set to 0). At this point, the master terminal waits for the slave device to respond with an ACK1 signal (the master sends a response from the device to send ACK1=0). Once the master device receives an ACK1 signal from the slave, the master sends a register address to the slave and waits for the its response to an ACK2 signal. When the master device receives ACK2, the master sends 8-bit data to be written, and then the master terminal waits for the slave device to respond with an ACK3 signal again. If the responses of the three ACK1~ACK3 signals from slave device are all zero, it indicates that the signal transmitted from the master to the slave device is normal.

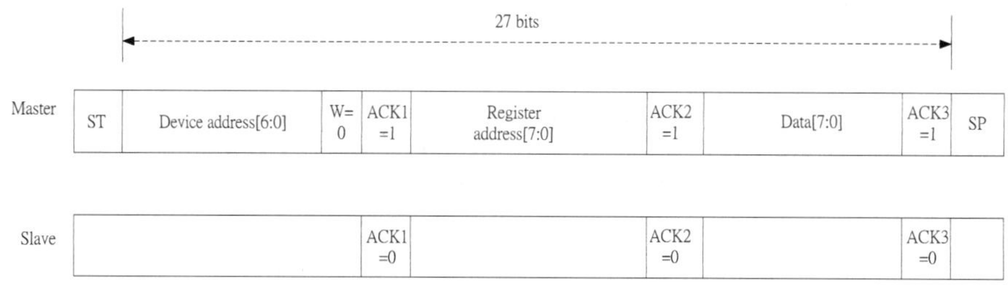

Figure 3. 8-bit I2C write operation bitmap.

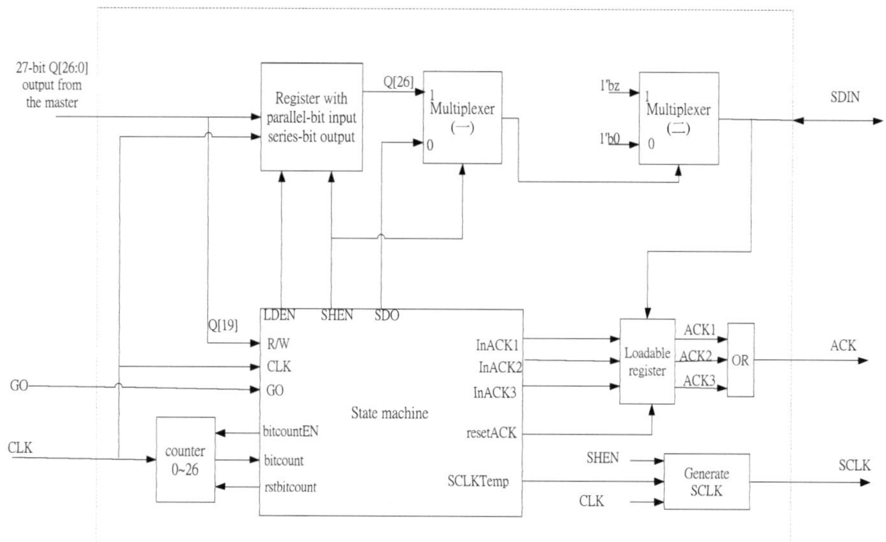

Figure 4. Master I2C write diagram.

Finally, the master sends the end signal SP to the slave one. (The interpretation of the ST and SP signals is shown in Figure 2). The above description is the process of writing data to the registered address of the slave address (Device address [6:0]). The transmitting bit diagram is shown in Figure 3. The I2C of master writing program is referred to conference [1] which block diagram is shown in Figure 4. The function of each sub-block is described as follows:

(i) 「Counter」: Because the master control element in Figure 3 has 27 bits to pass to the slave device, there must be a counter from 0 to 26 to count the current bit transmission. We design the counter by Simulink, as in Figure 5. After the simulation result is correct, the graphical program can be directly converted to the corresponding Verilog language as shown in Figure 6. There is no need to manually rewrite the Verilog language, which significantly saves time in programming the Verilog language.

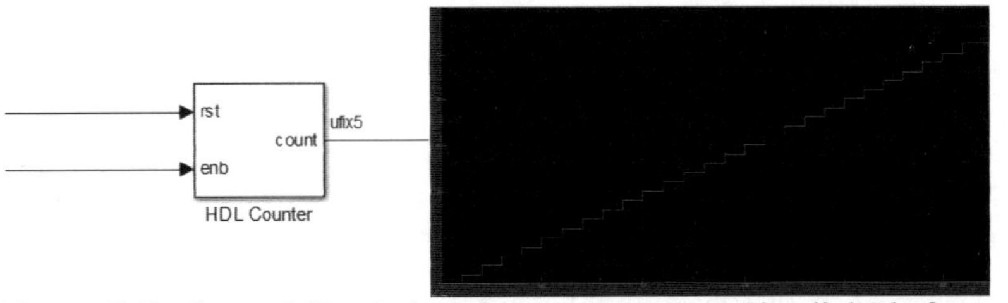

Figure 5. Design and Simulation of 0~16 counters on Simulink Platform.

```
// ---------------------------------------------

`timescale 1 ns / 1 ns

module HDL_counter
          (
           clk,
           reset,
           clk_enable,
           enab,
           reset_1,
           ce_out,
           count
          );


  input   clk;
  input   reset;
  input   clk_enable;
  input   enab;
  input   reset_1;
  output  ce_out;
  output  [4:0] count;  // ufix5


  wire enb;
  wire [7:0] HDL_Counter_out1;  // uint8
  reg [7:0] HDL_Counter_step_reg;  // ufix8
  wire [7:0] HDL_Counter_stepreg;  // uint8
  reg [7:0] HDL_Counter_count;  // ufix8
  wire [4:0] Data_Type_Conversion_out1;  // ufix5


  assign enb = clk_enable;

  // Count limited, Unsigned Counter
  //  initial value   = 0
  //  step value      = 1
  //  count to value  = 26
  always @(posedge clk or posedge reset)
    begin : HDL_Counter_step_process
      if (reset == 1'b1) begin
        HDL_Counter_step_reg <= 8'b00000001;
```

Figure 6. Verilog program of 0~16 counter converted by Simulink program Figure 5.

(ii) 「Shift register with parallel-bit input/serial single-bit output」: The function of this block is left-shift action per clock. That is, The parallel input is a 27-bit character and output is a single bit output. The input 27th bit is output with left rotation in every clock time. The Simulink platform design and simulation are shown in Figure 7.

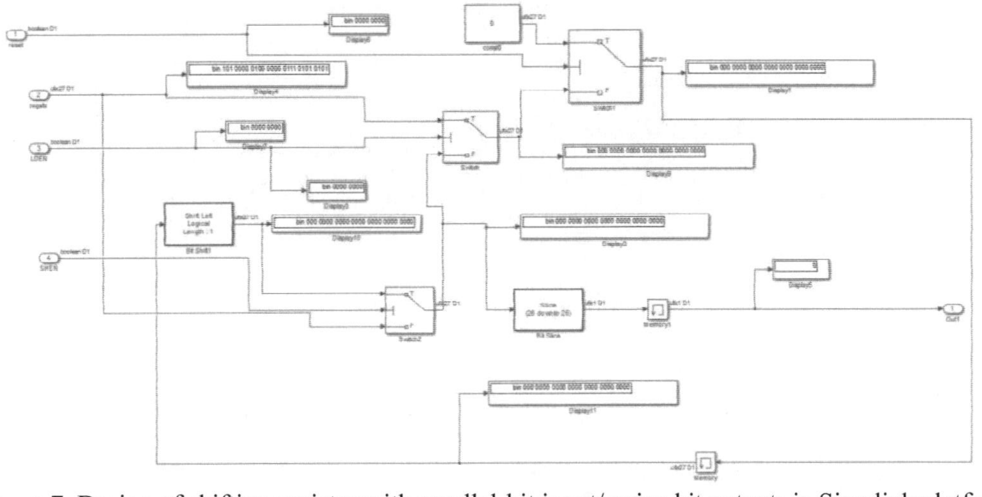

Figure 7. Design of shifting register with parallel-bit input/series-bit outputs in Simulink platform.

(iii) 「State machine for I2C writing」: The master of the I2C communication writes the characters of the state of each clock of the slave device shown in Figure 4. The states of state-machine are X_IDLE, X_GO, X_START, X_WAIT, X_SHIFT, X_STOP, X_FINAL, X_END which timing

IOP Conf. Series: Materials Science and Engineering **658** (2019) 012002 doi:10.1088/1757-899X/658/1/012002

diagrams are shown in Figure 8. The signal values corresponding to each state and the next state are summarized in Table 1. According to Table 1, we design the program on the Simulink platform as shown in Figure 9 which simulation waveform is shown in Figure 10.

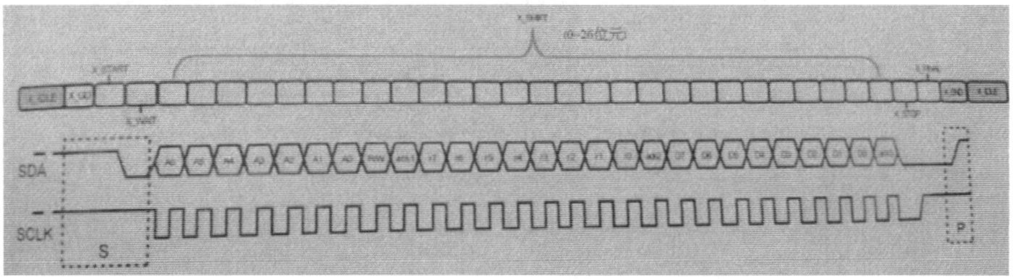

Figure 8. The write of state machine timing diagram of the master I2C [8].

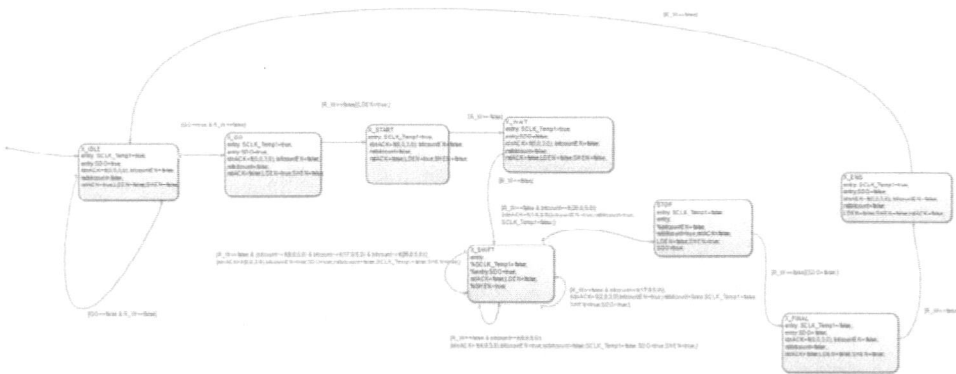

Figure 9. Design of the state machine of I2C write in the Simulink platform.

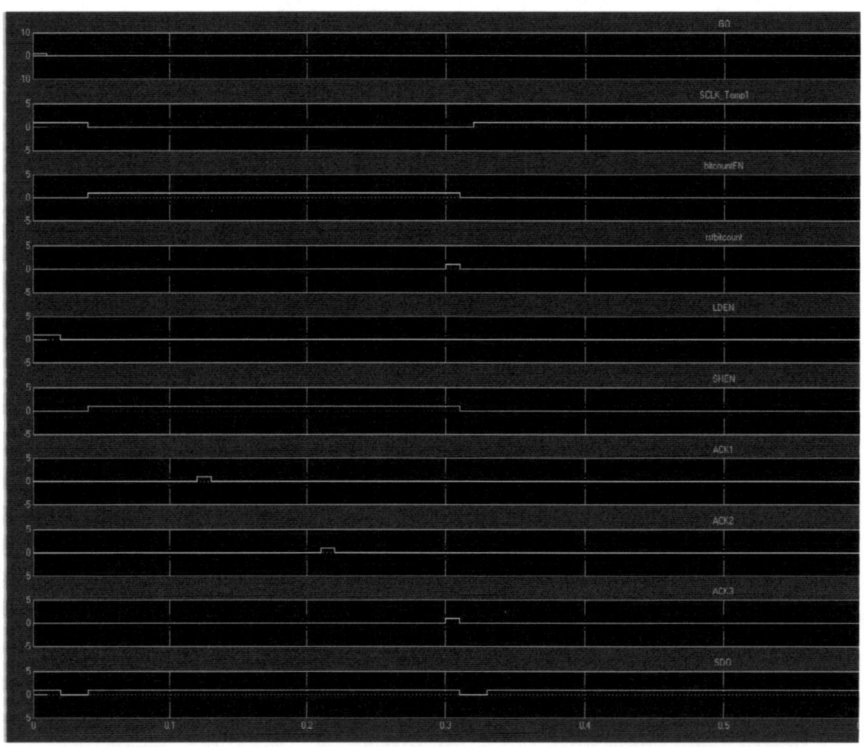

Figure 10. Simulation of the state machine of I2C writes for Figure 9.

IOP Conf. Series: Materials Science and Engineering **658** (2019) 012002 doi:10.1088/1757-899X/658/1/012002

(iv) 「Loadable Registers」:If reset=1, the output ACK1=0, ACK2=0, ACK3=0, Otherwise if ACK1=0, ACK2=0, ACK3=0 if resetACK=1, otherwise ACK1=SDIN if InACK1=1, ACK2=SDIN if InACK2=1, ACK3=SDIN if InACK1=3, In this way, it can be confirmed whether the slave device has received the signal from the master device. If so, ACK=0 in the block of Figure 4. This block of Simulink simulation is shown in Figure 11.

(v) 「Generate SCLK」: The relationship between the three inputs (CLK, SHEN, SCLK_Temp) and one output SCLK is when SHEN is greater than or equal to 1 SCLK=-CLK, otherwise SCLK=SCLK_Temp. The Simulink program is shown in Figure 12. After the Simulink program simulation of I2C master write diagram Figure 4 confirms that the simulation results are correct, the Simulink program can be directly converted into Verilog language and then integrated into the Quartus II software platform shown as Figure 13 which is a successfully compiled program. This Quartus II, FPGA-specific software, project is a fully Verilog language program for I2C communication.

Table 1. input/output relationships between states of I2C write state machine.

	Input			Output										
Current state	Counter value	trigger	Read/Write	Output										Next state
state	bitcount[4:0]	GO	R_W	SCLK_Temp	SCLK	SDIN	SDO	{IdnACK1, IdnACK2, IdnACK3}	Bitcount EN	Rst bitcount	Rst ACK	LDEN	SHEN	Next State
X_IDLE	X	0	0(W)	1	1	1'b1	1	3'b000	0	0	0	0	0	X_IDLE
X_IDLE	X	1	0(W)	1	1	1'b1	1	3'b000	0	0	0	0	0	X_GO
X_GO	X	X	0(W)	1	1	1'b1	0	3'b000	0	0	0	1	0	X_START
X_START	X	X	0(W)	1	1	1'b0	0	3'b000	0	0	0	0	0	X_WAIT
X_WAIT	X	X	0(W)	1	1	1'b0	0	3'b000	0	0	0	0	0	X_SHIFT
X_SHIFT	Countingfor not 8 and not 17 and not 26)	X	0(W)	0	~CLK	Q[26]	1	3'b000	1	0	0	0	1	X_SHIFT
X_SHIFT	8	X	0(W)	0	~CLK	1'bz	1	3'b100	1	0	0	0	1	X_SHIFT
X_SHIFT	17	X	0(W)	0	~CLK	1'bz	1	3'b010	1	0	0	0	1	X_SHIFT
X_SHIFT	26	X	0(W)	0	~CLK	1'bz	1	3'b001	1	1	0	0	1	X_STOP
X_STOP	X	X	0(W)	0	1'b0	1'b0	0	3'b000	0	0	0	0	0	X_FINAL
X_FINAL	X	X	0(W)	1	1'b1	1'b0	0	3'b000	0	0	0	0	0	X_END
X_END	X	X	0(W)	1	1'b1	1'b1	1	3'b000	0	0	1	0	0	X_IDLE

IOP Conf. Series: Materials Science and Engineering **658** (2019) 012002 doi:10.1088/1757-899X/658/1/012002

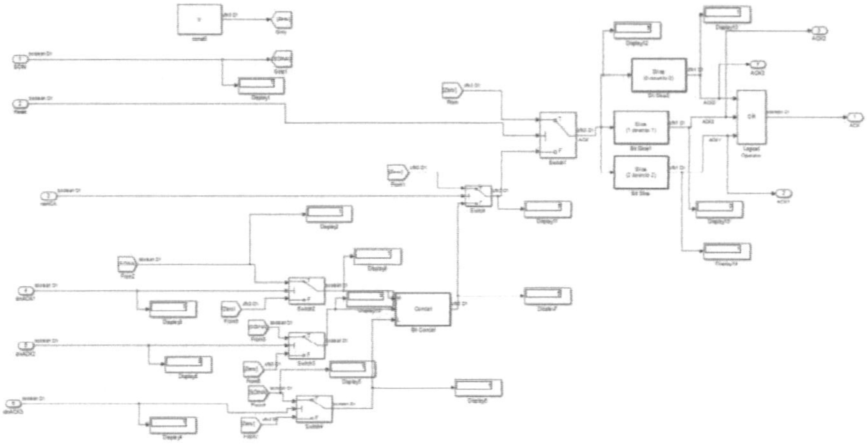

Figure 11. Loadable register design on Simulink platform.

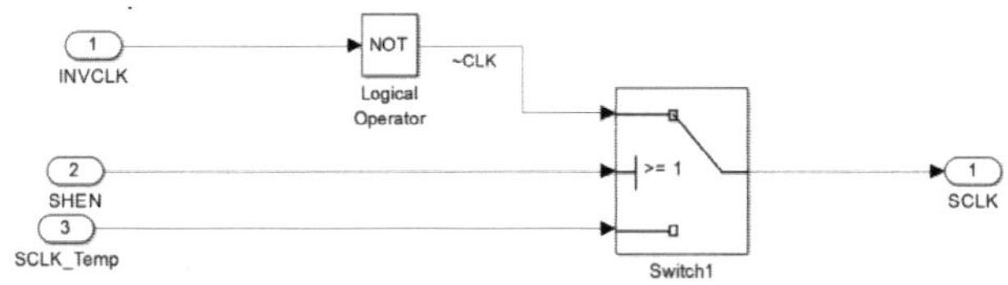

Figure 12. Simulink program of generating SCLK.

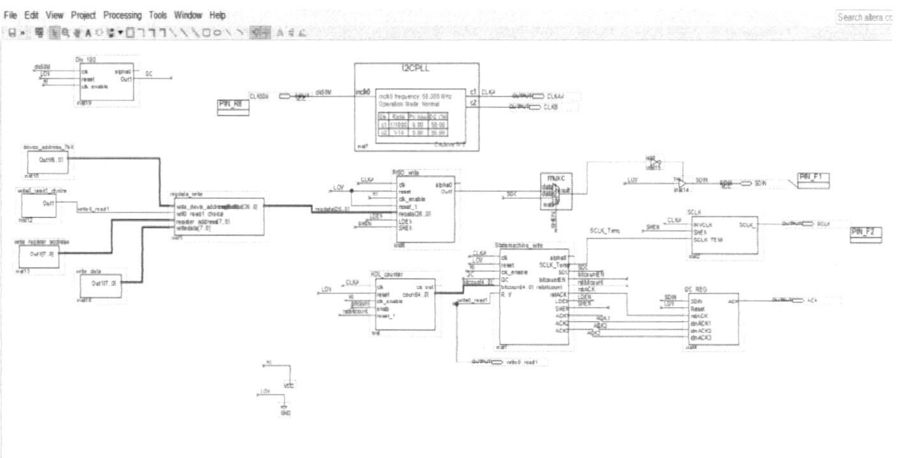

Figure 13. Corresponding Verilog language block integration of Figure 4.

4. I2C read action of the master device

Read an 8-bit data via the register address (Register address [7:0]) of the slave device. The control method is to generate a start signal ST from the master device first. Then, the master sends a set of address 7 bits (Device address [6:0]) plus a write (W) 1 bit (write set to 0). At this point, the master waits for slave device to respond with an ACK1 signal (the master sends a response from the device to send ACK1=0). Then the master sends a register address [7:0] to the register and waits for the slave to respond with an ACK2 signal (the master sends 1 to send a response with ACK2=0 from the device).

The master sends 1 bit named "Sr". This bit represents the start signal (SCLK = 1 indicates that the SDIN should have a moving action from 1 to 0.). Then the master control again sends a set of 7 address bits (Device address[6:0]) to be controlled, plus one read (R) bit (read set to 1) and waits for the slave responds with an ACK3 signal (the slave response ACK3=0 when the master sends

ACK3=1). The master sends 8 bits 11111111 to the slave device, waits for the slave respond to these 8 bits (the 8-bit value to be read), and then the master sends NACK=1 to the slave device.

Finally the master sends an end signal of the SP to complete the reading of the data from the loadable register. Its transmission bitmap is shown in Figure 14. Based on the above discussion, this paper proposes a program for I2C master control reading. The block diagram is shown in Figure 15. The function of each sub-block is described as follows:

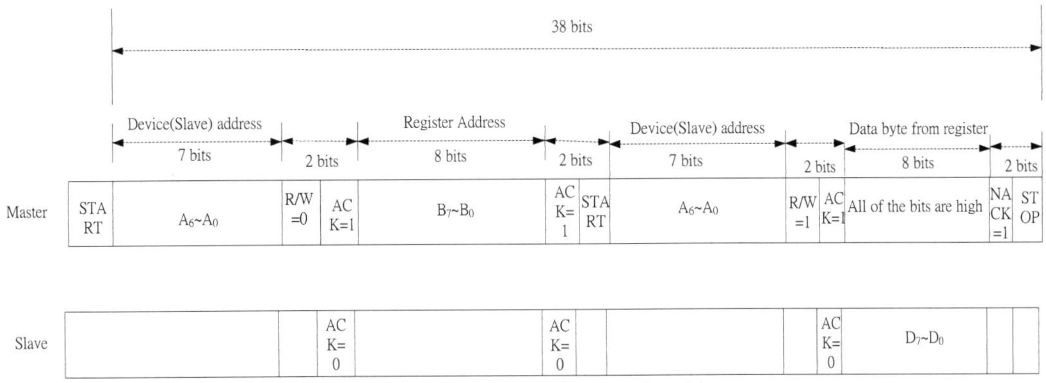

Figure 14. 8-bit I2C read operation bitmap.

(i) 「Counter」: Since there are 38 bits to be controlled, a counter from 0 to 37 is needed to design like the module in Figure 5 which is easy to program for tuning the inside parameters to obtain counter function.

(ii) 「Shift register with parallel bits input series bit output」: The input is a 38-bit character (0 to 37 characters). Each clock performs a left-shifting action, allowing only the 37th bit of the highest bit to pass through a single bit, which is the 37th-bit output per clock time. Its design diagram is similar to Figure 7, the number of transmitting bits is 38.

(iii) 「The state machine of I2C read」: Take use of the state machine to design the master of I2C communication that reads the state of the slave device data. Similar to Figure 8 shows the timing chart. The status of the reading of the master is divided into 8 states, X_IDLE, X_GO, X_START, X_WAIT, X_SHIFT, X_STOP, X_FINAL, X_END. The corresponding values of the states and the next state are summarized in Table 2. Complete the state machine design of Table 2 in the Simulink platform shown as Figure 16.

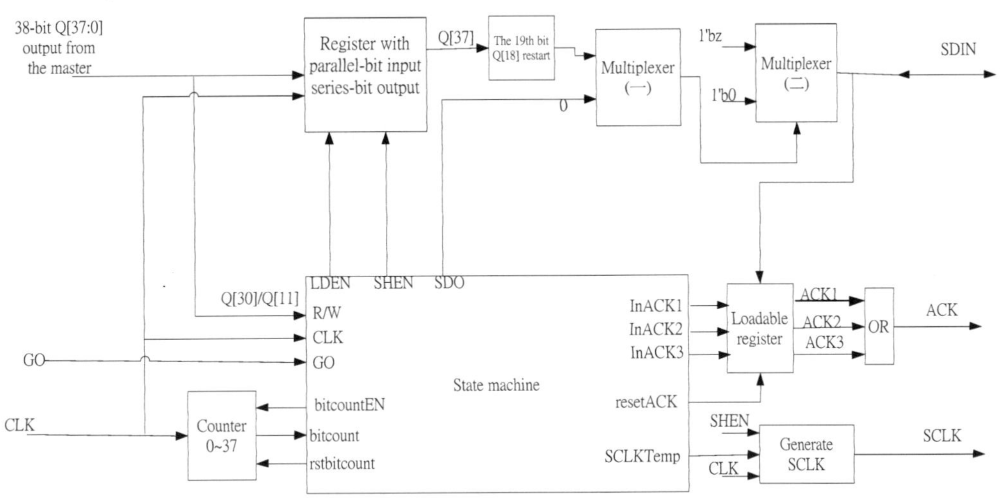

Figure 15. The master I2C read diagram.

(iv) 「The 19th bit Q[18] indicates restart」:When count through the counter (from 0 to 37) to 19, we let a clock with a phase shift of 90 degrees be treated as a single bit Q[18] so that is a restart action. The remaining modules in Figure 15 all have the same function as the module written by the master I2C in Figure 4.

The master I2C block Figure 15 is also simulated with Simulink programming. After verifying simulation, they can be converted into Verilog language and integrated into Quartus II software as shown in Figure 17. Projects in the Quartus II software are I2C-read communications programs.

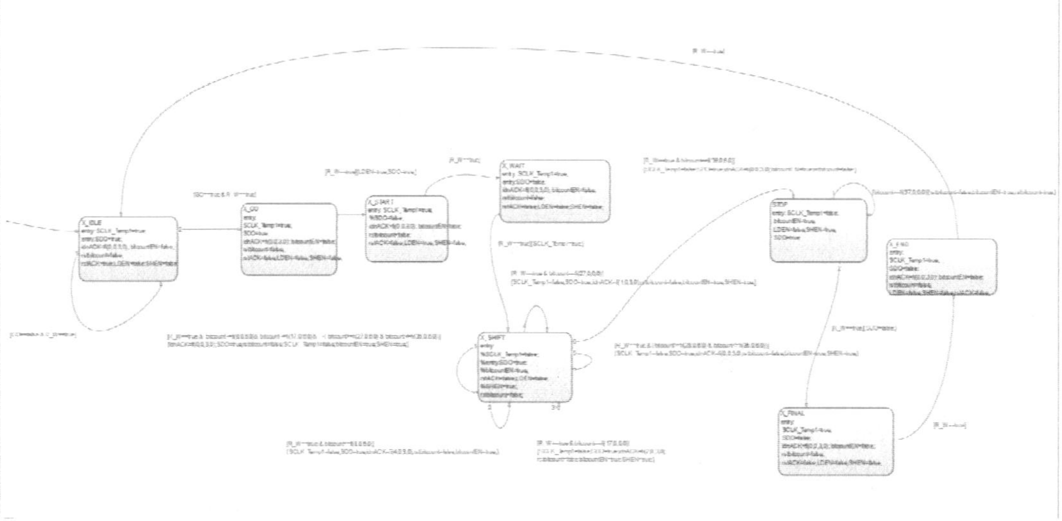

Figure 16. Design of the I2C of read state machine on the Simulink platform.

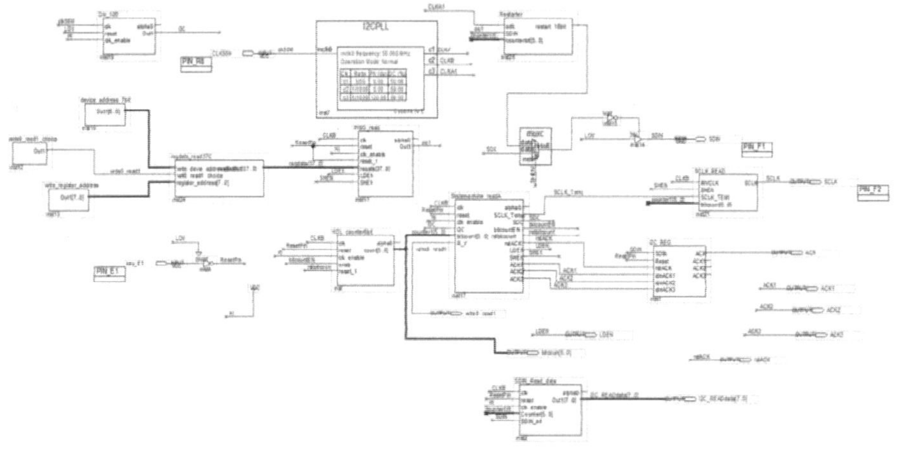

Figure 17. The corresponding Verilog block integration of Figure 15.

Table 2. input/output relationships between states of I2C read state machine.

	Input			Output										
Current state	Counter value	Trigger	Read/Write	output										Next state
state	bitcount[5:0]	GO	R_W	SCLK_Temp	SCLK	SDIN	SDO	{IdnACK1, IdnACK2, IdnACK3}	Bitcount EN	Rst bitcount	RstACK	LDEN	SHEN	Next State
X_IDLE	X	0	1(R)	1	1	1'b1	1	3'b000	0	0	0	0	0	X_IDLE

X_IDLE	X	1	1(R)	1	1	1'b1	1	3'b000	0	0	0	0	0	X_GO
X_GO	X	X	1(R)	1	1	1'b1	0	3'b000	0	0	0	1	0	X_START
X_START	X	X	1(R)	1	1	1'b0	0	3'b000	0	0	0	0	0	X_WAIT
X_WAIT	X	X	1(R)	1	1	1'b0	0	3'b000	0	0	0	0	0	X_SHIFT
X_SHIFT	Start counting for not 8 and not 17 and not 27-36	X	1(R)	0	~CLK	Q[36]	1	3'b000	1	0	0	0	1	X_SHIFT
X_SHIFT	8	X	1(R)	0	~CLK	1'bz	1	3'b100	1	0	0	0	1	X_SHIFT
X_SHIFT	17	X	1(R)	0	~CLK	1'bz	1	3'b010	1	0	0	0	1	X_SHIFT
X_SHIFT	27	X	1(R)	0	~CLK	1'bz	1	3'b001	1	0	0	0	1	X_SHIFT
X_SHIFT	28-35	X	1(R)	0	~CLK	1'bz	1	3'b000	1	0	0	0	1	X_SHIFT
X_SHIFT	36	X	1(R)	0	~CLK	Q[36]	1	3'b000	1	0	0	0	1	X_STOP
X_STOP	37	X	1(R)	0	~CLK	Q[37]	1	3'b000	1	1	0	0	1	X_STOP
X_STOP	X	X	1(R)	0	1'b0	1'b0	0	3'b000	0	0	0	0	0	X_FINAL
X_FINAL	X	X	1(R)	1	1'b1	1'b0	0	3'b000	0	0	0	0	0	X_END
X_END	X	X	1(R)	1	1'b1	1'b1	1	3'b000	0	0	1	0	0	X_IDLE

Figure 18. DE0-Nano FPGA verification platform.

5. The actual circuit verification

In this paper, the DEO-Nano verification platform as shown in Figure 18. The Cyclone IV FPGA chip and the 24AA02 EEPROM (Electronically Erasable Read-Only Memory) whose element address is 1010000. The data written to EEPROM temporary address one is $(170)_{10}$=$(10101010)_2$. The master device transmits data with 27-bits [26:0] which binary expression is $(101000001000000011101010101)_2$. The I2C writing simulation is shown in Figure 19. From ACK=0 in Figure 19, the data sending from the master to the slave device is successful. That is, the master-side FPGA chip has completed the I2C write operation.

The reading program from FPGA to I2C is shown in Figure 17. We want the data previously written into the EEPROM be read out. The simulation is as shown in Figure 20. ACK = 0 indicates that the slave did receive control of the I2C read issued by the master, and also actually read the value 170 just written from the EEPROM from the signal probe Figure 20. From the simulation of Figures 19 and 20, it can be concluded that the architectural design is feasible.

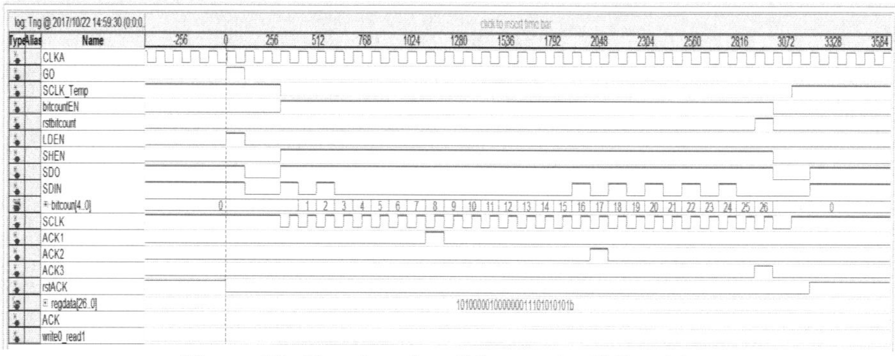

Figure 19. Signal probe of the master I2C writing.

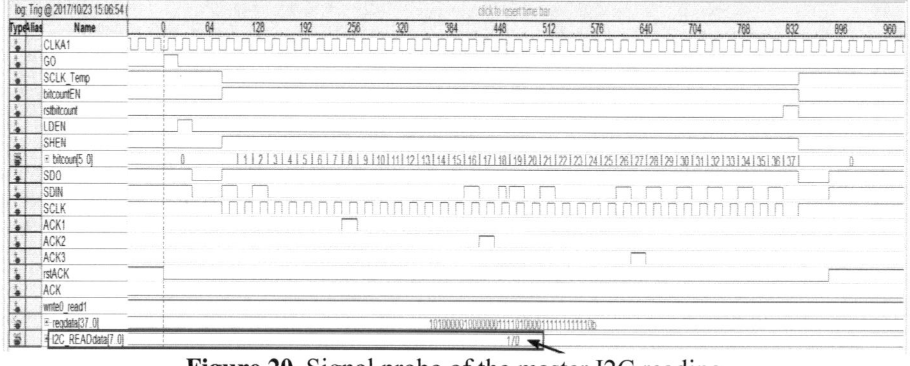

Figure 20. Signal probe of the master I2C reading.

6. Conclusions

This paper proposes an I2C write and read of architectural design for the FPGA master device. We also successfully completes its functional verification. The special point of this article is that it is not written directly in the Verilog language. Instead, all the sub-block functions are verified using the Simulink program. Then turn into the Verilog language that FPGA needs. Finally, use a logic analyzer to confirm sub-block function right or wrong. This design process is based on the Simulink block language. This is a way for engineers to shorten the writing time of the Verilog language. It is also a simulation and verification done at once, and there is a faster and clearer design step for the I2C design.

References

[1] Microchip Technology Inc. 2000, The I2C Bus Specification Version 2.1 https://www.csd.uoc.gr/~hy428/reading/i2c_spec.pdf

[2] Stephen B and Fischer R L 2000 An I2C Network Protocol for Environmental Monitoring, AN736 http://ww1.microchip.com/downloads/en/AppNotes/00736a.pdf

[3] Venkata Y. and Bin W. 2016 *Predictive Control of Wind Energy Conversion Systems* (John Wiley & Sons, Inc., Hoboken, New Jersey)

[4] Bai H, Liu C and Rathore A K 2019 *IEEE Transactions on Industrial Electronics* **66** pp 6581 – 6591

[5] Ana Dalia P A, Brisbane O M Esteban T C, Jesus Manuel M P and Luis G F 2019 *Communications In Nonlinear Science And Numerical Simulation* **72** pp 516 — 527
[6] Chen Z, He N and Huang Y 2019 *Genomics Proteomics & Bioinformatics* **16** pp 451 — 459
[7] MATLAB/SIMULINK TOOL, https://cn.mathworks.com/products/matlab.html
[8] Logical Circuit Design DE2-115 Practical Collection, Terasic Technology Publishing

An adaptive single channel EMD-TNMF blind source separation algorithm for both instantaneous and convolutive mixed signal

Pengju He[1,2], Mi Qi[2], Gangyi Liu[2], Zijiang Yu[2] and Qiang Fu[3]

[1]Research & Development Institute of Northwestern Polytechnical University in Shenzhen, China
[2]School of Automation, Northwestern Polytechnical University, Xi'an, China
[3]College of Aeronautics Engineering, Air Force Engineering University, Xi'an, China

E-mail: hepengju@nwpu.edu.cn

Abstract. Single channel blind source separation (SCBSS) using time-frequency nonnegative matrix factorization (NMF) has some shortcomings, where the source number and convolution order must be known and it's sensitive to the factors (e.g. window function, window length and overlap ratio of adjacent windows). To compensate these shortcomings that cannot meet the real application, we proposed an adaptive EMD-TNMF algorithm, which can estimate both the source number and the convolution order and is applicable to both single channel linear instantaneous and convolutive mixtures. Firstly, the single channel signal is mapped into multiple channels by utilizing empirical mode decomposition (EMD), and the number of independent sources is estimated by applying eigenvalue descent ratio of the IMFs' covariance matrix. Secondly, the autocorrelation-based method is used to estimate the convolution order. Thirdly, the nonnegative matrix is constructed by adding one positive matrix (all elements are positive). Finally, in time domain, NMF algorithm is used to separate source signals. The algorithm's performance is verified by two experiments where the single channel signal is linear instantaneous mixing of four artificial signals and linear convolutive mixing of two speech signals respectively. Results show that this algorithm can estimate the source number and convolution order correctly and obtain better separated source signals.

1. Introduction

In recent years, single channel blind source separation (SCBSS), which is an extreme case of underdetermined blind source separation (BSS), has become a research hotspot in signal processing. SCBSS is a novel technique for separating source signals from single channel mixed signal, which has broad application prospects. In the field of acoustics, it is necessary to separate multichannel speech signals from the observed signal which is collected by single sensor [1, 2]. In image processing, fingerprints obtained at the scene of criminal investigation are usually overlapped by multiple fingerprints, which should be solved by SCBSS algorithms [3]. In the field of biomedicine, interference signals (ECG, EMG, eye movement, etc.) have to be removed from single channel EEG signal [4, 5]. In fault detection, independent fault signals need to be separated from the observed signals for subsequent processing such as feature extraction and state recognition [6].

SCBSS has attracted much attention and relevant scholars have proposed many kinds of methods and techniques. In the early stage, pulse correlation adaptive-filtering, Fourier transform and wavelet transform are used to solve SCBSS problem, where the sources are only overlapped in time domain. Essentially, this kind of methods is transform domain filtering. J.R. Hopgood [7] discussed the problem of separability in detail and constructed separable transform domain. A. Ozerov [8] proposed to use time-frequency filter to actualize source separation, where short time Fourier transform (STFT) was employed. Though it can solve time domain overlap primely, it cannot process what are still overlapped in the transform domain. In transform domain methods, many parameters should be known in advance or could be estimated, and the accuracy of which greatly affected separation performance. Later, the researchers proposed to implement SCBSS in time or frequency domain by using the idea of characteristic bases or primary function [9, 10]. G.J. Jang [9] put forward using time domain primary functions to encode the source signals in a statistically efficient manner. M.J. Reyes-Gomez [11] used the basis function method to establish the HMM model. But it is necessary to know the true source signals in advance for this kind of model-based blind source separation methods, which limits its practical application. In the recent years, the research on the SCBSS problem mostly combines multichannel mapping with traditional blind separation methods such as independent component analysis (ICA) [12] and NMF [13, 14]. To solve the SCBSS problem of convolutive mixing, most scholars transform convolutive mixing into instantaneous mixing by employing time-frequency transformation [15-18], and then carry out instantaneous blind source separation algorithm. D. Kounades-Bastian [19] proposed a probabilistic framework based on NMF model and local complex Gaussian model in time-frequency domain. By Modeling time-varying hybrid filters via continuous time stochastic processes and using Kalman smoothing to estimate the time-varying mixed matrix and Wiener filtering to estimate the sources respectively. It solved the problem of time-varying convolutive mixtures of audio sources. However, the algorithm is complex and it assumes that the number of source signals is known and requires a higher degree of priori parameters.

To sum up, we find that the current SCBSS methods have the following shortcomings: (1) most methods are applicable to the case that contains only two source signals. In practice, the sources collected by a single sensor may be more than two; (2) the number of sources is known and remains unchanged, but it may vary in practice. For example, the number of sources will change with the sensor installation position in mechanical fault detection; (3) convolutive mixed signals need to be transformed into time-frequency domain, which increases the computational complexity and running time and needs to determine the phase of sources. In practical application, it is unable to guarantee the real-time performance and the hardware implementation is more difficult; (4) algorithms for linear instantaneous and convolutive mixing are proposed respectively, but there is no generic method that can solve both cases simultaneously.

In this paper, we propose an adaptive solution and utilize the EMD_TNMF algorithm to solve the linear mixed SCBSS problem, which is applicable to both instantaneous and convolutive mixed problem, and this method avoids time frequency domain conversion. In this method, Multichannel mapping is used to solve the problem of two or more source signals. First, EMD is applied to decompose a single channel mixed signal into multiple IMFs. Second, the number of independent sources is estimated by eigenvalue method based on the covariance matrix of all IMFs. Third, using the rank of sample autocorrelation matrix to estimate the order of hybrid filter. And then the nonnegative matrix is constructed according to the number of independent sources and IMFs. Finally, the TNMF is used to realize the blind source separation. The experimental results show that the number of independent sources and the order of hybrid filter can be estimated correctly, and the sources can be separated correctly in time domain.

The remaining of this paper is organized as follows. Section 2 introduces the related work and the contributions of this article. The mathematical model of SCBSS and some assumptions are described in Section 3. EMD, eigenvalue-based source number estimation method and autocorrelation-based model order estimation method are described in Section 4. The BSS algorithm based on time domain nonnegative matrix factorization is introduced in Section 5. The experimental validation is reported in Section 6. Conclusions and future works are discussed in Section 7.

2. Related work

At present, multichannel mapping methods mainly include Wavelet [20], EMD [21, 22] and variational modal decomposition (VMD) [23, 24], and phase space reconstruction [25, 26]. However, wavelet decomposition needs to set wavelet bases and hierarchies artificially, and embedding dimension and time delay should be set for phase space reconstruction. In the cases of lacking prior knowledge, subjective selection greatly interferes with the separation results and the adaptability is poor. Z.J. Zhao [20] used the wavelet packet decomposition method to reconstruct signals, which are combined with the observed signal to form multichannel signals. Finally, the separated signals are obtained by Fastica algorithm. B. Gao [27] used EMD to decompose single channel signal and used STFT to implement time-frequency transformation, each frequency band is separated by sparse 2-D nonnegative matrix factorization (vSNMF2D) and then is transformed back into time domain by the inverse STFT, finally the separated sources are obtained by K-means clustering. In this method, clustering is used to sort the separated results of each frequency band, but the amplitude inconsistency is not considered and the clustering center needs to be designated artificially. Y.N. Guo [28] proposed using local mean decomposition (LMD) and independent component analysis (ICA) to decompose single channel signal into a series of Product Functions and obtain the independent sources respectively. But for low signal-to-noise ratio, the separation results are not satisfactory. Z.l. Wu [24] used VMD algorithm instead of EMD to transform the single channel into virtual multiple channels, so that underdetermined ill-conditioned problem can be transformed into a positive definite general problem.

The contributions of this paper can be summarized as follows: (1) the determination of filter order in linear mixed model, in consideration of the equivalence of convolutive mixed model and FIR system, we utilized the rank of sample autocorrelation matrix to determine the filter order. (2) aiming at the technical difficulty of sources number estimation, we combined EMD with eigenvalue-based method to estimate the number of independent sources in single channel mixture. (3) to solve the problem that corresponding algorithms are necessary for linear instantaneous and convolutive mixtures respectively, we combined the order estimation of linear mixing model and TNMF. The proposed algorithm is applicable to those two kinds of linear mixing SCBSS problems, avoiding the time-frequency domain conversion for convolutive mixtures and reducing complexity and uncertainty.

3. Linear mixed model of SCBSS

In the practical observation environment, source signals propagate through various paths. Due to the time delays of different propagation paths, the signals observed by sensors are not only instantaneous aliasing of source signals, but also convolutive aliasing of multiple source signals at current and past times. Therefore, BSS should extend to convolutive mixtures. In this paper, we studied two cases of single channel linear mixing, namely instantaneous and convolutive mixing.

a. Single channel linear instantaneous mixed model

Assuming that the observed single channel mixed signal $\boldsymbol{x}(t)$ is a linear instantaneous mixture of r independent source signals. Its mathematical model [29] can be expressed as:

$$\boldsymbol{x}(t)=\sum_{i=1}^{r}a_i\boldsymbol{s}_i(t)+\varepsilon(t) \tag{1}$$

Where $s_i(t), i=1,2,\ldots,r$ is the mutually independent source signals, $\varepsilon(t)$ is white gaussian noise with mean of zero and variance of σ^2, a_i is the mixing coefficient and r is number of sources.

b. Single channel linear convolutive mixed model

Assuming that r mutually independent source signals are propagated by different paths and are perceived by one sensor. Considering time delays, the mixed signal $\boldsymbol{x}^*(t)$ is related to the current time and the preceding K-1 sample times of the source signals. Its mathematical model [30] can be expressed as:

$$x^*(t) = \sum_{i=1}^{r}\sum_{\tau=0}^{K-1} a_i(\tau)s_i(t-\tau) + \varepsilon(t) \tag{2}$$

Where $x^*(t)$ is the single channel convolutive mixed signal, $a_i(\tau)$ is the FIR filter coefficients between ith source and the sensor. $s_i(t-\tau),(i=1,2,...,r;\tau=0,1,...,K-1)$ are time delay components of source signals, K is the order of FIR filter and it is instantaneous mixing if and only if K=1. $\varepsilon(t)$ and r are the same as the above.

4. Technology and method

4.1. Multichannel mapping based on EMD

EMD is an adaptive signal decomposition method based on local characteristics of signals [31]. The basic idea of EMD is that any complex signal can be decomposed into components of simple non-sinusoidal functions which are different from each other. It can decompose the signal into a series of IMFs adaptively. The IMF component must satisfy two conditions: (1) at any point, the mean value of the envelope defined by its local maxima and local minima must be zero; (2) in the entire data sequence, the number of extrema and the number of zero-crossings must be equal or differ at most by one.

By employing EMD to decompose the single channel observed signal, multiple IMFs with different frequency components can be obtained. Each IMF is regarded as observed signal to realize multichannel mapping. Supposing that the observed signal is $x(t)$, the algorithm flow of EMD is shown in Table 1.

Table 1. EMD algorithm flow

1. Determine all extreme points of $x(t)$;
2.The upper envelope $x_u(t)$ and the lower envelope $x_l(t)$ are fitted by the three-spline function. Calculate the average of envelopes $m_1^1(t)=(x_u(t)+x_l(t))/2$ and $h_1^1(t)=x(t)-m_1^1(t)$;
3. If $h_1^1(t)$ satisfies the IMF conditions, then $c_1(t)=h_1^1(t)$. Otherwise taking $h_1^1(t)$ as object and repeat 1~2 to screen $h_1^k(t)=h_1^{k-1}(t)-m_1^{k-1}(t)$ until the IMF condition is satisfied, $c_1(t)=h_1^k(t)$;
4. $r_1(t)=x(t)-c_1(t)$, $x(t)=r_1(t)$;
5. repeat steps 1~4 to get the next IMF until the residue $r_n(t)$ is a monotonic function that can no longer be decomposed. Finally get $X_{imf}(t)=(c_1(t),c_2(t),...,c_n(t),r_n(t))^T$.

4.2. Estimation of source number and construction of nonnegative matrix

It is the prerequisite for single channel blind source separation to estimate the independent source number accurately. In convolutive mixed environment, the signal will produce mutation, which makes it difficult to estimate the number of sources. FFT-based source number estimation method [32] is no longer applicable. Therefore, we combined EMD and eigenvalue-based method [33, 34] to estimate the number of independent sources in single channel mixture.

The eigenvalue-based method estimates the number of sources through subspace analysis. Firstly, the singular value decomposition (SVD) is used to calculate the eigenvalues of covariance matrix of the mixed signals, and then the source number is determined according to the eigenvalue information. The eigenvalues are only affected by the noise intensity not by the mixing mode and the EMD method can eliminate the noise effect to a certain extent. Therefore, the eigenvalue-based method can estimate the number of independent sources both in instantaneous mixed model and in convolutive mixed model. The algorithm of source number estimation based on EMD and eigenvalue-based method is shown in Table 2.

Table 2. Source number estimation algorithm flow

1. Obtain $X_{imf}(t)=(c_1(t),c_2(t),...,c_n(t),r_n(t))^T$, use EMD to decompose single channel signal;
2. Compute correlation matrix of IMFs $R_{imf}=E\left[X_{imf}(t)X_{imf}(t)^T\right]$;
3. Singular value decomposition of R_{imf} and get eigenvalues $\lambda_1\geq\lambda_2\geq...\geq\lambda_n\geq 0$;
4. Calculate $k_1=\lambda_1/\lambda_2,k_2=\lambda_2/\lambda_3,...,k_{n-1}=\lambda_{n-1}/\lambda_n$, the descent ratio of adjacent eigenvalues;
5. Choose the largest drop ratio $k_r=\max\{k_i,i=1,2,...,n-1\}$, the source number is r.

After having estimated the number of independent sources, the multichannel signals are reconstructed by the method proposed in [35]. The reconstructed multichannel signals $x_{con}(t)$ are as follows:

$$x_{con}(t)=(c_1(t),c_2(t),...,c_{r-1}(t),c_r(t)+c_{r+1}(t)+...+c_n(t)+r_n(t))^T \tag{3}$$

For convolutive mixed signals, most scholars transform the single channel signal or the reconstructed signals into frequency domain and then apply instantaneous blind separation algorithm, such as nonnegative matrix factorization algorithm in time-frequency domain. This kind of method can not only solve the problem of nonnegativity of signals, but also turn convolutive mixture into instantaneous mixture. However, in the frequency domain, phase information is involved and the problems of sorting and classification must be solved in the frequency band. If instantaneous mixing and convolutive mixing can be solved directly in time domain, the complexity of the algorithm will be reduced and the algorithm will have more practical value.

Then how to make the reconstructed multichannel signals satisfy the nonnegativity in time domain? Firstly, determine the minimum negative number in the reconstructed multichannel signals, and then construct a positive matrix $|\min negative|*ones(r,l)$, where r is source number, l is the data length. By adding the positive matrix to $x_{con}(t)$, get the nonnegative matrix $x_{con\_nn}(t)$:

$$x_{con\_nn}(t)=x_{con}(t)+|\min negative|*ones(r,l) \tag{4}$$

4.3. Order estimation of linear convolutive mixtures

Inspired by literature [36, 37], the method based on MA model order estimation is proposed to determine the order of hybrid filters in SCBSS model. The system equation of moving average (MA) model is as follows:

$$x(t)=\theta_0 a_t-\theta_1 a_{t-1}-...-\theta_q a_{t-q}=\sum_{\tau=0}^{q}\theta_\tau a_{t-\tau} \tag{5}$$

Comparing linear mixed models (1) and (2) with the MA model, both models can be interpreted as an output sequence driven by an input sequence to a linear system, which are equivalent in terms of system and model. Therefore, the filter order of the linear mixing model in formula (2) can be estimated by determining the order of the MA model of the output sequence. For an ARMA process, the problem of determining the order of MA (filter order K in this paper) is simple in theory, but the effectiveness of the actual algorithm is difficult to solve. J. Chow[38] has proved that the MA order q of the ARMA process can be determined by formula(6), the order q is the largest integer l that makes (6) establishment. This formula is established only at $l=q$, and for $0<l<q$, it is not completely satisfied. In order to get the largest integer, statistical tests must be carried out, but no specific algorithm has been introduced.

$$R_x(l)+\sum_{i=1}^{p} a_i R_x(l-1) \neq 0 \quad (l \leq q) \tag{6}$$

X.D. Zhang [39] proposed a numerical robust SVD method to estimate the order of MA model, selecting the autocorrelation matrix with $q_e > q$ dimensions, which is closely related to Hankel matrix. The method starts with $Q = q_e > q$ and reduces one dimension in turn, using SVD to determine the rank of $\boldsymbol{R}_{1e}$. When $Q = q$, it is the first time that the rank changes from p to $p+1$ (when $q_e > q$, $rank(R_{1e}) = p$, and if only if $q_e = q$, $rank(R_{1e}) = p+1$).

$$\boldsymbol{R}_{1e} = \begin{bmatrix} R_x(q_e) & R_x(q_e-1) & \cdots & R_x(q_e-p) \\ R_x(q_e+1) & R_x(q_e) & \cdots & R_x(q_e-p+1) \\ \vdots & \vdots & \ddots & \vdots \\ R_x(q_e+p) & R_x(q_e+p-1) & \cdots & R_x(q_e) \end{bmatrix} \tag{7}$$

Because we used the sample autocorrelation, the turning point is not obvious. Using the "overdetermined" matrix $\boldsymbol{R}_{2e}[i,j] = R_x(q_e - p + i + j - 2), i = 1,2,...,t; j = 1,...,p_e+1, t \gg p_e$, obviously it satisfies $rank(\boldsymbol{R}_{2e}) = rank(\boldsymbol{R}_{1e})$. The filter order estimation algorithm is shown in Table.

Table 3. Filter order estimation algorithm flow

1. Set or estimate the AR order of the model and $Q = q_e > q$;
2. Construct sample correlation matrix $\boldsymbol{R}_{2e}$ and SVD of $\boldsymbol{R}_{2e}$;
3. $Q \leftarrow Q-1$, construct sample correlation matrix $\boldsymbol{R}_{2e}$ and SVD of $\boldsymbol{R}_{2e}$;
4.If there is an obvious turning point of the $p+1$ singular value between this time and last time, $q = Q$; conversely, repeat the steps 3~4 until q is selected.

Using the ratio α in formula (8) as the transition detection criteria,

$$\alpha = \frac{\sigma_{p+1,p+1}^{(Q)} - \sigma_{p+1,p+1}^{(Q+1)}}{\sigma_{p+1,p+1}^{(Q+1)}} \tag{8}$$

Where $\sigma_{p+1,p+1}^{(Q)}$ and $\sigma_{p+1,p+1}^{(Q+1)}$ are the $(p+1)th$ singular value when Q and Q+1 respectively. If the relative change rate α is greater than the given threshold value, then accept this Q value as the turning point, that is, the filter order is K=Q.

5. Time domain nonnegative matrix factorization

D.D. Lee and H.S. Seung [40] proposed nonnegative matrix factorization, its nonnegative constraints make the physical meaning of the decomposition result clear, and it is consistent with the nonnegative data in speech, image, etc. This method is widely used in speech separation, image compression and other fields. The basic NMF method can be applied to the over-determined blind source separation problem of linear instantaneous mixtures, but it cannot be directly applied to the separation of convolutive mixtures in time domain due to the time dependence of signals. P. Smaragdis [13] proposed a nonnegative matrix factorization deconvolution method, the signal is converted to time-frequency domain by STFT, and the amplitude spectrum is selected as the input to extract multiple independent sources. In this paper, the reconstructed signal matrix satisfies non-negativity and there is no need to use transform domain method. In time domain, each independent source signal can be extracted effectively through the nonnegative matrix factorization method by considering time correlation of signals.

Given a nonnegative matrix $\boldsymbol{V} \in R^{m \times n}$, find a set of matrices $\boldsymbol{W}(\tau) \in R^{m \times r} (\tau = 1,2,...,K-1)$ and $\boldsymbol{H} \in R^{r \times n}$, where r is the rank of decomposed matrix and $r \leq \min(m,n)$, such that:

$$V = \sum_{\tau=0}^{K-1} W(\tau) \overset{\tau\rightarrow}{H} \qquad s.t. W(\tau), H \geq 0 \tag{9}$$

Where V represents the reconstructed nonnegative matrix, K is the filter order, $W(\tau)$ is the encoding matrix and H is the basis matrix. If applying this to blind source separation, the element $w_{ij}(\tau)$ of $W(\tau)$ means the τth coefficient of the FIR filter between the jth source signal and the ith mixed signal, $\overset{\tau\rightarrow}{H}$ means τ time delay component of source signal, the operator shifts the columns of its argument by τ spots to the right. So that:

$$H = \begin{bmatrix} h_{11} & \cdots & h_{1(n-1)} & h_{1n} \\ \vdots & \ddots & \vdots & \vdots \\ h_{r1} & \cdots & h_{r(n-1)} & h_{rn} \end{bmatrix}, \overset{1\rightarrow}{H} = \begin{bmatrix} 0 & h_{11} & \cdots & h_{1(n-1)} \\ \vdots & \vdots & \ddots & \vdots \\ 0 & h_{r1} & \cdots & h_{r(n-1)} \end{bmatrix}, \ldots \tag{10}$$

In convolutive mixed model, $X(t) = [x_1(t), x_2(t), ..., x_r(t)]^T$ represents the r channel reconstructed convolutive mixed signals, obtained by multichannel mapping of the single channel mixed signal, which is the K order convolutive mixing of r-channel source signals $S(t) = [S_1(t), S_2(t), ..., S_r(t)]^T$. Written in matrix form is as follows:

$$\begin{bmatrix} x_1(t) \\ x_2(t) \\ \vdots \\ x_r(t) \end{bmatrix} = \begin{bmatrix} a_{11} & a_{12} & \cdots & a_{1r} \\ a_{21} & a_{22} & \cdots & a_{2r} \\ \vdots & \vdots & \vdots & \vdots \\ a_{r1} & a_{r2} & \cdots & a_{rr} \end{bmatrix} \begin{bmatrix} S_1(t) \\ S_2(t) \\ \vdots \\ S_r(t) \end{bmatrix} \tag{11}$$

Where matrix element a_{ij} is a vector that $a_{ij} = [a_{ij}(0), a_{ij}(1), \cdots, a_{ij}(K-1)](i, j = 1, \cdots, n)$, it requires linear convolution between a_{ij} and each source to get the observed signals. As shown in the following formula:

$$X = \sum_{\tau=0}^{K-1} A(\tau) \overset{\tau\rightarrow}{S} \tag{12}$$

Comparing formulas (2) and (12), we find that $W(\tau)$ and $H = [h_1(t), h_2(t), \cdots, h_n(t)]^T$ are the estimation of mixing matrix $A(\tau)$ and source signals respectively. Using Kullback-Leibler divergence as the distance measure [41], the optimization problem of TNMF is with the following form:

$$\begin{gathered} E = \left\| V \odot \ln\left(\frac{V}{\hat{V}}\right) - V + \hat{V} \right\| \\ \hat{V} = \sum_{\tau=0}^{K-1} W(\tau) \overset{\tau\rightarrow}{H} \end{gathered} \tag{13}$$

During each iteration process, updating $W(\tau)$ and H for every τ. Focusing on the above optimization problem, gradient descent method is used to update $W(\tau)$ and H iteratively. The update rules for TNMF are as follows:

$$\begin{cases} H \leftarrow H \odot \dfrac{W(\tau)^T \cdot \overset{\leftarrow\tau}{\left[\dfrac{V}{\hat{V}}\right]}}{W(\tau)^T \cdot ones} \\ W(\tau) \leftarrow W(\tau) \odot \dfrac{\dfrac{V}{\hat{V}} \cdot \overset{\tau\rightarrow}{H}^T}{ones \cdot \overset{\tau\rightarrow}{H}^T} \end{cases} \tag{14}$$

Where 'ones' serves as $m \times n$ matrix that all elements are 1, $\odot$ is the Hadamard product. According to update rules, iteration is carried out until E in formula (13) is less than a certain threshold. When it is instantaneous mixing, the elements a_{ij} of the mixing matrix is scalar. Namely when the filter order is K=1, formula (9) can be simplified to $\boldsymbol{V} = \boldsymbol{WH},\ s.t.\boldsymbol{W},\boldsymbol{H} \geq 0$. Correspondingly, formula (12) is simplified to $\boldsymbol{X} = \boldsymbol{AS}$. Linear instantaneous mixing is the special case of linear convolutive mixing. Therefore, TNMF can be used as a general method to solve the blind separation problem of instantaneous mixing and convolutive mixing in time domain. In summary, the specific steps of the proposed SCBSS algorithm based on EMD-TNMF are shown in Table 4.

Table 4. SCBSS algorithm flow based on EMD-TNMF

1. Single channel observed signal $\boldsymbol{x}(t)$ in formula (1) or $\boldsymbol{x}^*(t)$ in formula (2);
2. Use EMD in 4.1 to decompose single channel observed signal into multiple IMF components, $\boldsymbol{X}_{imf}(t) = (\boldsymbol{c}_1(t),\boldsymbol{c}_2(t),...,\boldsymbol{c}_n(t),\boldsymbol{r}_n(t))^T$;
3. Use the eigenvalue-based method in 4.2 to estimate the source number r;
4. According to formulas (3) and (4), get the nonnegative matrix $\boldsymbol{x}_{con\_nn}(t)$;
5. Use the filter order estimation method in 4.3 to estimate the order K of linear mixing;
6. Get the independent sources $s_i(t), i = 1,\cdots,r$ by TNMF in Section 5.

6. Experiment and result

In this section, two groups of experiments are established, we select artificial signals and speech signals respectively to verify the effectiveness and accuracy of the proposed algorithm. It is proved that this algorithm can achieve great results both for linear instantaneous mixed and linear convolutive mixed SCBSS problem. This algorithm is compared with the Wavelet-TNMF algorithm and EMD-vSNMF2D algorithm.

6.1. Establish experiments

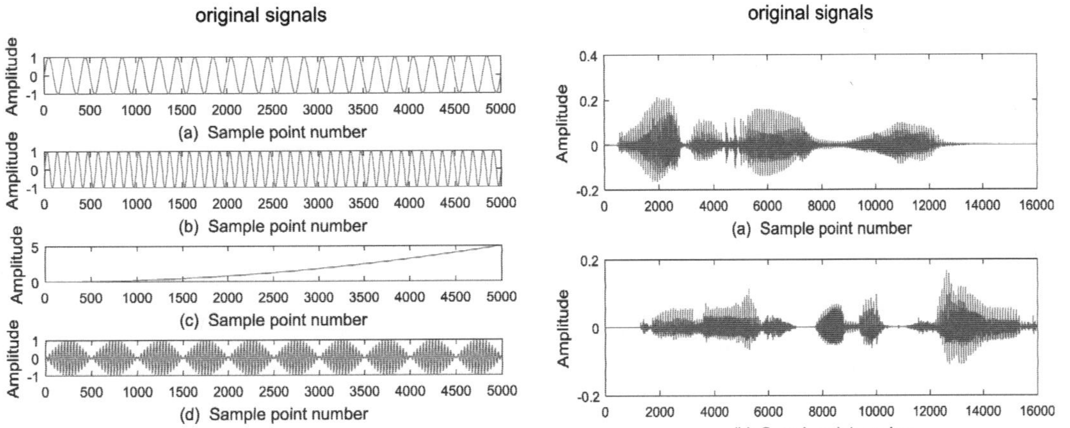

(a) (b)

Figure 1. (a) Four channels of artificial signals: two sinusoidal signals with different frequencies, a quadratic curve signal and a modulation signal. (b) Two channels of speech signals are selected from the TIMIT standard speech database.

(1) Four channel artificial signals(as shown in Figure 1(a)) are two sinusoidal signals with different frequencies $s_1 = \sin(2\pi 25t / fs)$ and $s_2 = \sin(2\pi 45t / fs)$, a quadratic curve signal $s_3 = 5 \times (t / fs).\wedge 2$ and one modulation signal $s_4 = \sin(2\pi 5t / fs).*\sin(2\pi 150t / fs)$ respectively. The sampling frequency is fs=5000Hz and the sample points are t=1:5000. This set of experimental data is used to verify that the algorithm can separate single channel linear instantaneous mixture with more than two sources.

(2) Source signals are two speech signals from the TIMIT standard speech database, sampling frequency *fs*=16KHz, the unit of signal level is *mV* and sample points are 16000(as shown in Figure 1(b)). This set of experimental data is used to verify the separation of single channel convolutive mixture.

The simulation platform of the experiments is MATLAB R2016a and runs under the Windows 7 operating system of Intel (R) Core (TM) i5-3230M CPU@2.60GHz and 8GB RAM.

6.2. Experimental verification

6.2.1. SCBSS of instantaneous mixing with more than two sources. Single channel observed signal: using the four channel artificial signals in 6.1 (1) as source signals and generate a random 1×4 mixed matrix $[0.3229 \quad 0.3282 \quad 0.7909 \quad 0.4875]$. The source signals are mixed according to Formula (1) and then add the white gaussian noise of 10dB, single channel observed signal is shown in Figure 2.

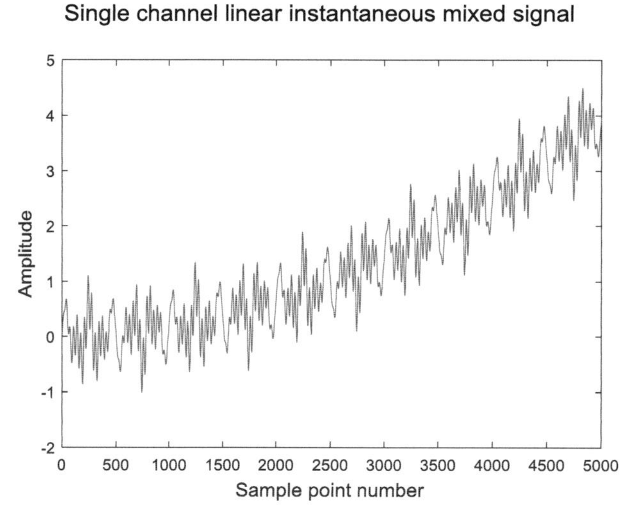

Figure 2. Single channel linear instantaneous mixed signal.

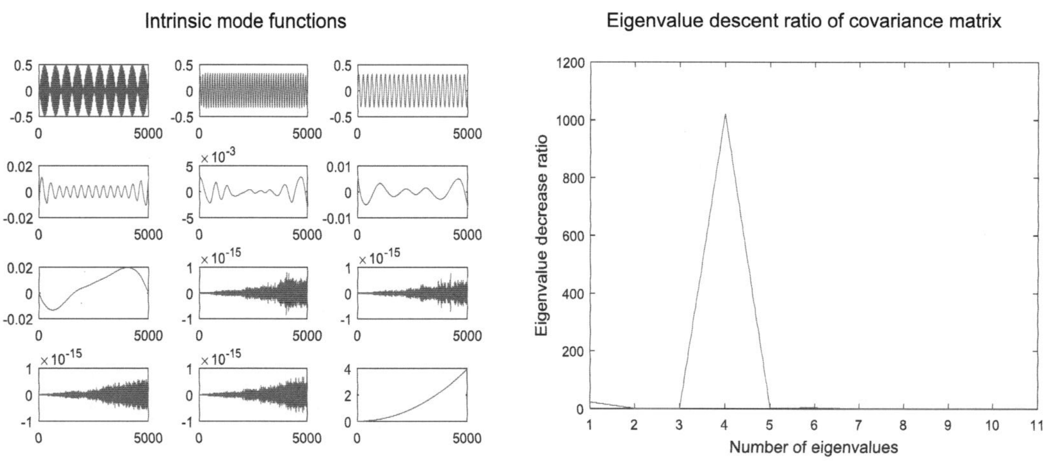

(a) (b)

Figure 3. (a) The IMFs obtained by empirical mode decomposition. From left to right, from top to bottom, according to the order of high frequency to low frequency. The last one is residual component or trend item. (b) The descent rate of the eigenvalue of the covariance matrix. When n=4, the rate of eigenvalue descent is the largest, so the number of source signals is 4.

Multichannel mapping and source number estimation: implement EMD and obtain a set of IMFs, as shown in Figure 3(a). The eigenvalues of the covariance matrix of IMFs are calculated by SVD. The descent ratio of the eigenvalue of the covariance matrix is shown in Figure 3(b). The number of independent sources estimated by eigenvalue-based method is 4.

Filter order estimation (sample points are 1000): when Q=2, the singular values are $\sigma_{11}=1.5019\times10^3, \sigma_{22}=448.0948, \sigma_{33}=199.7901, \sigma_{44}=9.86, \sigma_{55}=0.019, \sigma_{66}=0.0171, \sigma_{77}=0.0087$, $\sigma_{88}=4.3803\times10^{-4}, ..., \sigma_{1515}=2.3215\times10^{-14}$.It can be seen that the first six singular values occupy the dominant position, and so p=6.When Q=1, the singular values are $\sigma_{11}=1.5800\times10^3, \sigma_{22}=451.3304$, $\sigma_{33}=196.77, \sigma_{44}=9.62, \sigma_{55}=0.0268, \sigma_{66}=0.0168, \sigma_{77}=0.0092, \sigma_{88}=0.0084, ..., \sigma_{1515}=2.9458\times10^{-14}$. At Q=1, the value of σ_{88} meets the requirements and there is an obvious turning point $\alpha=1817.68\%$. Therefore, the filter order is $K=q=1$. The filter order estimated by many times experiments is 1, that means the hybrid model is linear instantaneous mixing.

Nonnegative matrix construction and blind source separation: given the number of independent components and IMFs, according to formulas (3) and (4), the constructed nonnegative matrix is shown in Figure 4(a). After these steps, the number of source signals, filter order and reconstructed observation signals are obtained. The reconstructed signals are taken as input and implement time domain nonnegative matrix factorization, four separated signals are shown in Figure 4(b).

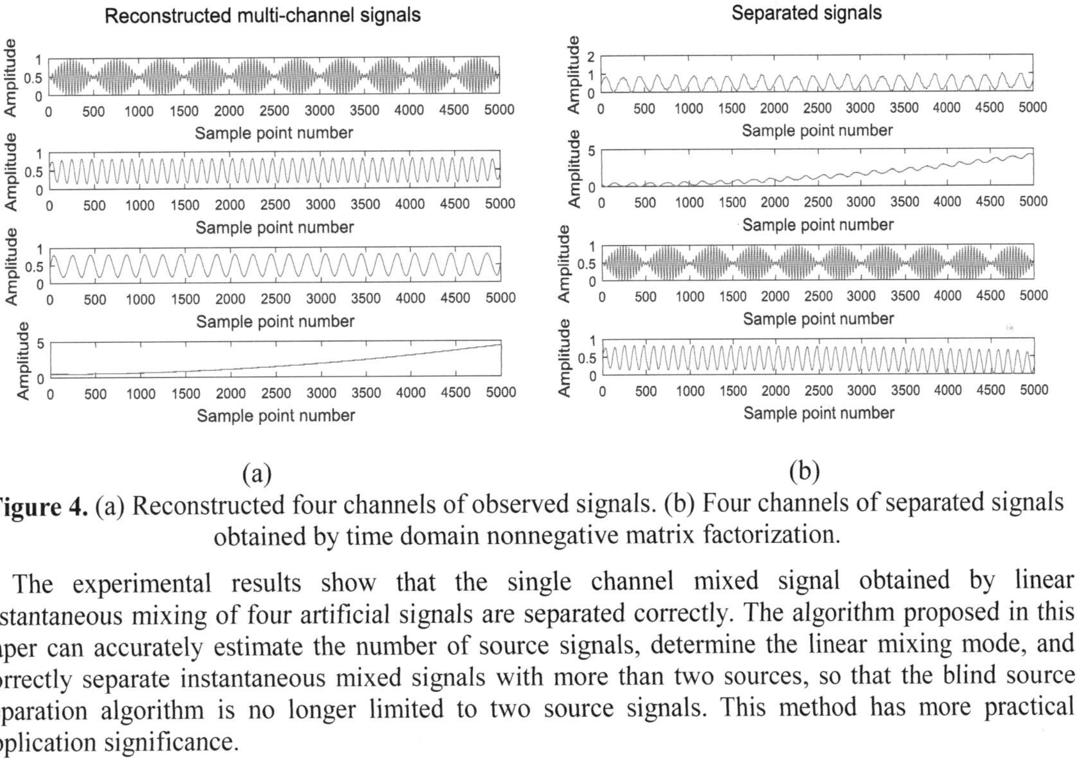

Figure 4. (a) Reconstructed four channels of observed signals. (b) Four channels of separated signals obtained by time domain nonnegative matrix factorization.

The experimental results show that the single channel mixed signal obtained by linear instantaneous mixing of four artificial signals are separated correctly. The algorithm proposed in this paper can accurately estimate the number of source signals, determine the linear mixing mode, and correctly separate instantaneous mixed signals with more than two sources, so that the blind source separation algorithm is no longer limited to two source signals. This method has more practical application significance.

6.2.2. SCBSS of convolutive mixing with two speech signals. The observed signal received by a single sensor is often the convolutive mixing of multi-channel source signals, due to the influence of different propagation paths, reflection and so on. This experiment simulates a single microphone to capture multiple speeches, and uses this algorithm to separate the two speech signals.

Single channel observed signal: using the two channel speeches in 6.1(2) as source signals, convolutive mixing matrix is $A=[a_{11} \quad a_{12}]$,where $a_{11}=[0.8450\ 0.4670\ 0.6834\ \ 0.2073\ 0.7440]$ and

$a_{12} = [0.2097\ \ 0.3685\ \ 0.0350\ \ 0.1809\ \ 0.6882]$. Generating single channel signal according to formula (2) and adding the white gaussian noise of 10dB, the single channel observed signal is shown in Figure 5.

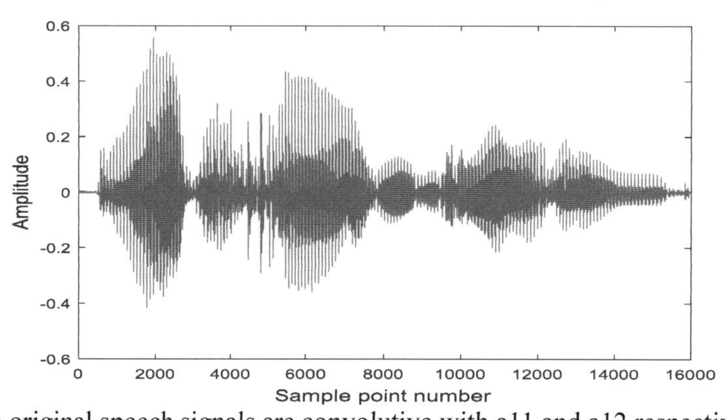

Figure 5. The two original speech signals are convolutive with a11 and a12 respectively, and then add them to obtain the single channel convolutive mixed signal.

Multichannel mapping and source number estimation: implement EMD and obtain a set of IMFs, as shown in Figure 6(a). The eigenvalue-based method is used to estimate the number of independent components, and the descent ratio of eigenvalues is shown in Figure 6(b). The number of independent components is 2.

Filter order estimation (sample points are 1000): when Q=6, the singular values are $\sigma_{11} = 9.2488,\ \sigma_{22} = 7.5419,\ \sigma_{33} = 1.1679,\ \sigma_{44} = 0.5739,\ \sigma_{55} = 0.0692,\ \sigma_{66} = 0.0046, \sigma_{77} = 7.8649\times10^{-4}, \ldots,$ $\sigma_{1515} = 2.5031\times10^{-5}$. It can be seen that the first six singular values occupy the dominant position, and so $p = 6$ .when Q=5, the singular values are： $\sigma_{11} = 9.8018,\ \sigma_{22} = 7.6088,\ \sigma_{33} = 1.1223,\ \sigma_{44} = 0.6024,$ $\sigma_{55} = 0.0687, \sigma_{66} = 0.0046, \sigma_{77} = 0.0018, \ldots,\ \sigma_{1515} = 2.5064\times10^{-5}$. At Q=5, the value of σ_{77} meets the requirements and there is a more obvious turning point $\alpha = 128.86\%$. Therefore, the filter order is $K = q = 5$. The filter order estimated by many times experiments are 5, that is, the mixing mode is linear convolutive mixing.

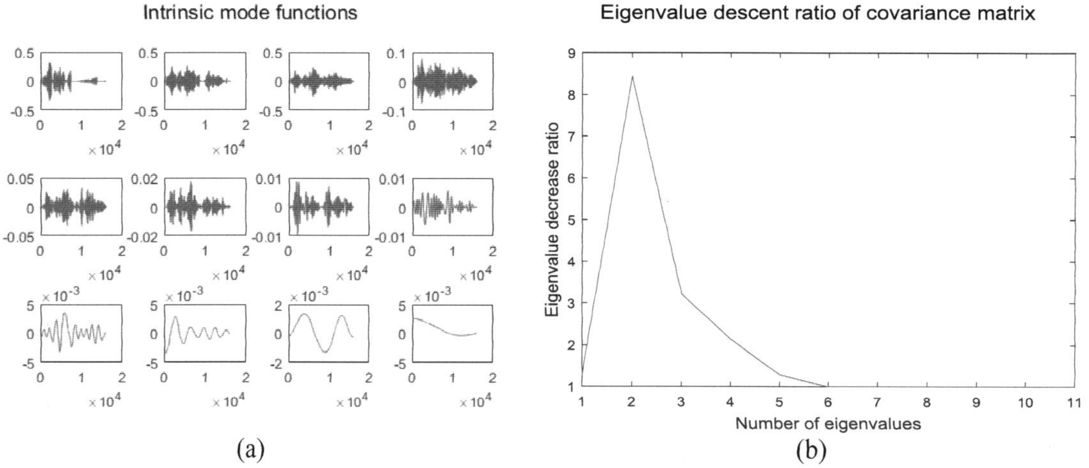

(a) (b)

Figure 6. (a) The IMFs obtained by empirical mode decomposition. From left to right, from top to bottom, according to the order of high frequency to low frequency. The last one is residual component or trend item. (b) The descent rate of the eigenvalue of the covariance matrix. When n=2, the rate of eigenvalue descent is the largest, so the number of source signals is 2.

Nonnegative matrix construction and blind source separation: given the number of independent components and IMFs, according to formulas (3) and (4), the constructed nonnegative matrix is shown in Figure 7(a). After these steps, the number of source signals, filter order and reconstructed observation signals are obtained. The reconstructed observed signals are taken as input and implement time domain nonnegative matrix factorization, two separation signals are shown in Figure 7(b).

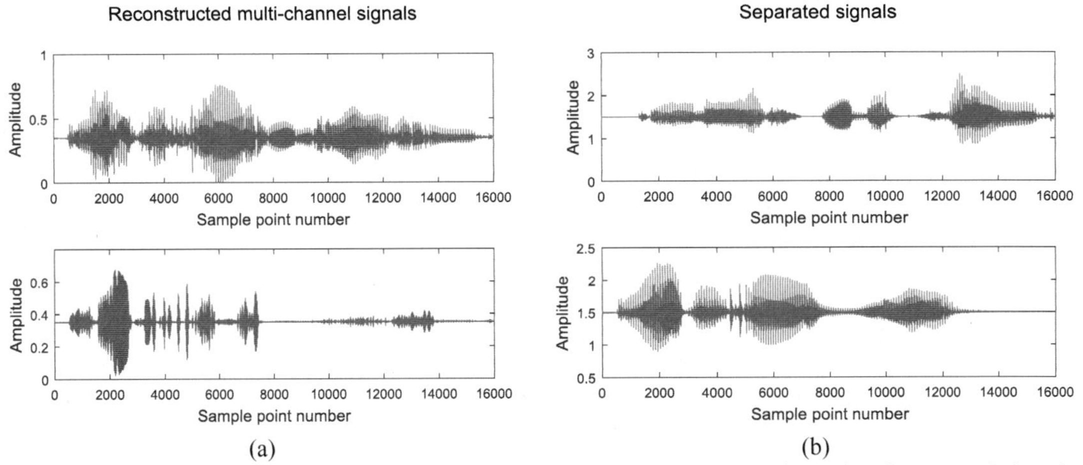

Figure 7. (a) Reconstructed two channels of observed signals. (b) Two channels of separated signals obtained by time domain nonnegative matrix factorization.

From the experiment of two speech signals, it can be concluded that the proposed algorithm can estimate the number of independent components in the single channel linear convolutive mixed signal, determine the linear mixing mode and correctly separate the source signals.

6.3. Performance evaluation

Signal to noise ratio (SNR) is used to be as a criterion for judging the separation effect of blind source separation. The formula is as follows:

$$SNR[\hat{s}_i(t)] = 10\log\frac{E[|s_i(t)|^2]}{E[|\hat{s}_i(t)-s_i(t)|^2]} \tag{15}$$

Where $s_i(t)$ is the true sources and $\hat{s}_i(t)$ is the estimated source. The higher the signal-to-noise ratio, the better the performance of the algorithm. However, the premise of this standard is to correctly estimate the power or amplitude of the signal. Therefore, choosing the modified version of (15) in[42]. Assuming that the projection on corresponding source signals of $\hat{s}_i(t)$ is $r_i s_i(t)$, in this way, SNR can be expressed as:

$$SNR[\hat{s}_i(t)] = 10\log\frac{E[|r_i s_i(t)|^2]}{E[|\hat{s}_i(t)-r_i s_i(t)|^2]} \tag{16}$$

Where the value of r_i is what makes $E[|\hat{s}_i(t)-r_i s_i(t)|^2]$ obtain the minimum value. It is not necessary that the amplitude of the separated signals is equal to that of the true source signals, avoiding the shortcomings of formula (15).

6.3.1. Comparison between EMD and Wavelet. For multichannel mapping problem of single channel, comparing EMD with wavelet transform in this paper. Different wavelet bases produce different results. Using s_1 and s_4 in 6.1 (1) as the source signals of contrast experiment. Three commonly used wavelet bases "Coiflets", "Daubechies" and "BiorSplines" are used to divide the single channel signal into 10 layers by wavelet transform, and two observed signals are reconstituted. The blind source separation algorithm TNMF proposed in this paper is used to achieve separation.

Using the improved SNR formula as the performance evaluation criterion, the separation results of the proposed algorithm and the Wavelet-NMF with three different wavelet bases are shown in Table 5.

Table 5. The SNR (dB) of the separated signals obtained by different multichannel mapping methods.

	EMD	Wavelet(coif)	Wavelet(db)	Wavelet(bior)
S1	13.48	12.57	9.43	11.71
S2	12.73	11.69	10.58	8.25

The experimental results show that the separated results of the algorithm in this paper is better than that of Wavelet-NMF. The choice of wavelet bases has an important influence on the separated results. The wavelet transform needs to set the number of layers and wavelet bases artificially, so its adaptability is poor. While the EMD not only has the advantage of adaptability, but also has better performance than Wavelet-NMF.

6.3.2. Comparison between TNMF and vSNMF2D Comparing a variety of nonnegative matrix methods:1. the method proposed in this paper EMD-TNMF; 2. Wavelet-TNMF, the wavelet base is "Coiflets"; 3. EMD-vSNMF2D. Choose two speech signals as the source signals of the experiment and obtain the single channel observed signal according to formula (2). The improved SNR formula is used to evaluate the performance of each algorithm. The comparison results are shown in Table 6. And the running time of the three methods is shown in Table 7.

Table 6. SNR (dB) comparison results of three algorithms

	EMD-TNMF	Wavelet-TNMF	EMD-vSNMF2D
Speech1	14.8	13.6	13.9
Speech2	13.5	13.2	12.4

Table 7. Running time of the three methods(/s)

Methods	EMD_TNMF	Wavelet-TNMF	EMD-vSNMF2D
Running time	72.36	75.65	285.78

The experimental results in Table 6 and Table 7 show that the EMD-TNMF is superior to the other two algorithms. The method EMD-vSNMF2D not only makes the data complex in frequency domain conversion, but also requires a designated center for clustering method, which increases the uncertainty of separated signals. Although the SNR and running time of EMD-TNMF are not much different from those of the Wavelet-NMF method, it has been proved in 6.3A that the Wavelet is less adaptive than EMD.

Typical artificial signals and speech signals were selected in the experiments. It is verified that the effectiveness and practicability of the single channel blind source separation algorithm proposed in this paper. The results show that the proposed algorithm EMD-TNMF can solve the blind source separation problem of single channel linear mixture with two and more than two sources. It can correctly estimate the number of independent sources, determine the linear mixing mode and can be applied to engineering practice.

7. Conclusions

In this paper, we proposed an adaptive EMD-TNMF algorithm for SCBSS, which is applicable to both single channel linear instantaneous and convolutive mixed signals. In this method, EMD is used to decompose the single channel signal into multiple IMFs, the number of independent components is estimated by eigenvalue method based on the covariance matrix of IMFs and the mixing mode is determined by autocorrelation-based filter order estimation method. The virtual multichannel signals are reconstructed and the nonnegative condition is satisfied by adding a positive matrix. Finally, source signals are separated by the TNMF algorithm in time domain. The feasibility of the algorithm is verified by synthetic signals and speech signals in the experiments. The results show that the algorithm can correctly estimate the number of independent sources and the order of hybrid filter, and realize the separation of two or more source signals for SCBSS. This paper provides a complete solution for linear mixed SCBSS problem.

References

[1] Mowlaee P, Saeidi R, Christensen M G and Martin R 2012 *IEEE Int. Conf. on Acoustics* pp 69−72
[2] Mehrez S, Keisuke K, Marc D and Tomohiro N 2014 *J. IEEE/ACM Transactions on Audio, Speech and Language Processing* **22**(2) pp 354−367
[3] Wei W, Jianwei L, Feifei H and Hailiang F 2008 *J. Pattern Recognition Letters* **29**(3) pp 301−308
[4] Guofa S and Lei D 2015 *J. Brain Topography* **28**(1) pp 47−61
[5] Pengju H, Xiaomeng C, Yan L and Huazhi Z 2016 *MATEC Web of Conf.* **44** 01026 pp 1−9
[6] Jiamin L, Junwei L and Ling P 2017 *J. Journal of Chongqing University* **40(7)** pp 25−31
[7] Hopgood J R and Rayner P J W 2003 *J. IEEE Transactions on Signal Processing* **51**(7) pp 1739−1752
[8] Ozerov A, Philippe P, Gribonval R and Bimbot F 2005 *IEEE Workshop on Applications of Signal Processing to Audio and Acoustics* pp 90−93
[9] Jang G J, Lee T W and Oh Y H 2003 *J. IEEE Signal Processing Letters* **10**(6) pp 168−171
[10] Gao B, Woo W L and Dlay S S 2008 *3rd Int. Conf. on Information and Communication Technologies: From Theory to Applications* pp 1−5
[11] M.J. Reyes-Gomez, D.P.W. Ellis and N. Jojic 2004 *IEEE Int. Conf. on Acoustics, Speech, and Signal Processing* **vol 5** pp 641−644
[12] Davies M E and James C J 2007 *J. Signal Processing* **87**(8) pp 1819−1832
[13] Smaragdis P 2004 *Lecture Notes in Computer Science* **3195** pp 494−499
[14] Smaragdis P, FéVotte C, Mysore G J, Mohammadiha N and Hoffman M 2014 *J. IEEE Signal Processing Magazine* **31**(3) pp 66−75
[15] Kadambe S and Boudreaux-Bartels G F 1992 *J. IEEE Transactions on Signal Processing* **40**(10) pp 2498−2517
[16] Hiroe A 2007 *ICA'07 Proc. of the 7th Int. Conf. on Independent component analysis and signal separation* pp 471−479
[17] Durak L and Arikan O 2008 *J. IEEE Transactions on Signal Processing* **51**(5) pp 1231−1242
[18] Lu W K and Zhang Q 2009 *J. IEEE Signal Processing Letters* **16**(7) pp 576−579
[19] Kounades-Bastian D, Girin L, Alameda-Pineda X, Gannot S, Horaud R and Abed K *2016 J. IEEE/ACM Transactions on Audio, Speech and Language Processing* **24**(8) pp 1408−1423
[20] Zhijing Z and Yanbo H 2017 *J. Communications Technology* **50**(3) pp 425−429
[21] Tsolis G S and Xenos T D 2011 *Int. Journal of Signal Processing Image Processing & Pattern Recognition* **4**(2) pp 91−106
[22] Guo Y, Huang S and Li Y 2012 *J. Circuits Systems & Signal Processing* **31**(6) pp 2047−2060
[23] Dragomiretskiy K and Zosso D 2014 *J. IEEE Transactions on Signal Processing* **62**(3) pp 531−544
[24] Zhongliang W and Min W 2018 *J. Communications Technology* **51**(4) pp 774−777
[25] Dai W, Huang D and Liu B 2015 *J. Empire Survey Review* **47**(345) pp 387−396
[26] Pengju H, Tingting S, Wenhui L and Weibiao Y 2017 *J. Mechanical Systems & Signal Processing* **113** pp 22−35
[27] Gao B, Woo W L and Dlay S S 2011 *J. IEEE Transactions on Audio Speech & Language Processing* **19**(4) pp 961−976
[28] Guo Y, Naik G R and Nguyen H 2013 *35th Annual Int. Conf. of the IEEE Engineering in*

Medicine and Biology Society (EMBC) pp 6812 — 6815

[29] Jang G J and Lee T W 2014 *J. Journal of Machine Learning Research* **4** pp 1365 — 1392

[30] Pankiewicz P J, Arildsen T and Larsen T 2013 *J. Mathematics* pp 1 — 10

[31] Huang N E, Shen Z, Long S R, Wu M C, Shih H H and Zheng Q 1998 *Proc. Mathematical Physical & Engineering Sciences* **454**(1971) pp 903 — 995

[32] Linh-Trung N, Belouchrani A, Abed-meraim K and Boashash B 2005 *J. EURASIP Journal on Applied Signal Processing* **2005** pp 2828 — 2847

[33] Lei J, Ping C and Juan Y 2007 *2nd IEEE Conf. on Industrial Electronics and Applications* pp 2727 — 2731

[34] Bin G 2011 *Single channel blind source separation* Newcastle University pp 1 — 155

[35] Pengju H and Xiaomeng C 2015 *J. Technology & Health Care* **24**(Suppl 1) pp 17 — 26

[36] Xianda Z and Yuansheng Z 1994 *J. IEEE Transactions on Signal Processing* **42**(10) pp 2854 — 2858

[37] Lee T W and Bell A J 1998 *J. Nips* **9** pp 758 — 764

[38] Chow J 1972 *J. IEEE Transactions on Automatic Control* **17**(5) pp 707 — 709

[39] Xianda Z and Yuansheng Z 2002 *J. IEEE Transactions on Signal Processing* **41**(6) pp 2277 — 2280

[40] Lee D D and Seung H S 1999 *J. NATURE* **401**(6755) pp 788 — 798

[41] Févotte C, Bertin N and Durrieu J L *J. Neural Computation* **21**(3) pp 793 — 825

[42] Xu M B and Zhu W Z 2002 *J. Journal of Hangzhou Dianzi University* **22**(3) pp 63 — 66

IOP Conf. Series: Materials Science and Engineering **658** (2019) 012004 doi:10.1088/1757-899X/658/1/012004

Gamification of an asynchronous HTML5-related competency-based guided learning system

Cheng Hsiu Li

Department of Information Management, National Taitung Junior College, Taiwan

E-mail: lcs@ntc.edu.tw

Abstract. Practice e-learning systems with traditional asynchronous drills are the most common assisted-learning tools for cognitive study, and they are often employed in instructional environments. Such systems provide learners with immediate feedback that enables them to develop cognitive abilities through constant learning. In this study, a novel, asynchronous HTML5-related gamified competency-based guided learning (CBGL) system was designed. The CBGL algorithm was embedded in the system to provide personalization for enhanced learning effects. This study used the core drivers of the Octalysis gamification framework to design gamified mechanisms that ameliorated the long and tedious CBGL processes. This study used the system satisfaction test to survey the flow experience of the participants and logged their learning behaviors for data mining. The results of system satisfaction analysis revealed that the system gamification generated a favorable flow experience among participants; they actively used the system to study and felt that the guided learning process was engaging. Results of the data mining analysis also revealed that the percentage of learners who completed the CBGL process rose from 58.3% to 79.2%. Therefore, the gamified mechanism helped participants overcome the boring and lengthy learning processes of the CBGL system.

1. Introduction

Digital learning is defined as any distance education model that applies digital learning resources and digital media [24]. Since 1990, the advancement of the Internet has resolved problems related to updating and delivering learning content. This improvement has led to digital learning that can occur without limits to time and location; consequently, the Internet has become an essential medium for digital learning [24]. Many modern digital learning systems are web-based. The development of Internet technology has also added more innovative elements to digital learning, surpassing conventional learning media such as CDs, TVs, or computers, and evolving toward the era of e-learning 4.0. The basis of human knowledge acquisition lies in cognitive learning. The primary mediums of cognitive learning are practice e-learning systems with asynchronous drills that utilize cognitive resources. These systems provide instant feedback and are commonly used for teaching cognitive learning through continuous training and learning processes. However, in asynchronous digital learning environments, other factors (e.g., a lack of teacher supervision and a boring learning process) often affect learners' learning motivation and result in a poor learning outcome. Strategies to overcome this problem are necessary to encourage the adoption of digital learning systems and enable the learning process to become more autonomous, thereby motivating learners to use such systems.

In the 1970s, academics determined that humans obtain cognitive abilities, such as knowledge and memory, on the basis of behaviorism theory [3]. Cognitive development is the first stage at which humans recognize the world around them. Humans continuously gain knowledge through cognitive stimuli, cognitive responses, and experience feedback (S–R–F). Digital drill and practice learning is a type of e-learning that integrates behaviorism, and it is usually used in the learning of cognitive abilities. The S–R association model is formed from continuous stimuli and responses and strengthened by feedback, as illustrated by the behaviorism model in Figure 1. The law of effect dictates that praise and reward provide positive reinforcement and that punishment restrains wrong behavior. The law of contiguity emphasizes immediate feedback, with events that occur together likely to become associated; and the law of exercise emphasizes continuous practice as a route to attain mastery. On the basis of these laws, applying the correct response in the shortest time through continuous exercise can be emphasized through curriculum design to strengthen the S–R association. In drill and practice e-learning systems, questions acts as stimuli and answers act as responses. When answers are correct, systems add points; conversely, when answers are wrong, systems deduct points (as in the law of effect). The reward and punishment come from the immediate feedback (as in the law of contiguity). Moreover, learners using such systems must continuously practice (as in the law of exercise). Consequently, drill and practice e-learning systems can be useful tools for accelerating the S–R–F process during the learning of cognitive knowledge. However, because such systems emphasize continuous practice to increase efficacy, they are likely to cause boredom among students, and they do not motivate learners' or encourage self-study [19].

Competency-based guided learning (CBGL) systems are modified from training-based digital learning systems. The CBGL algorithm is incorporated into training-based systems, and the system emphasizes student-centered learning processes and individual differences among learners. Because cognitive learning is mostly evaluated using a limited item bank, if the learning content has a standard scope that can be classified into different units, CBGL systems can personalize learning paths and guide students to independently learn in stages according to the cognitive level of different units. The results of one study showed that students' learning outcomes could be significantly improved using a CBGL system [14]. The stimulation and construction of the learning environment directly affect students' learning. In digital learning environments, students' active attitudes affect their learning effectiveness [2]. However, the CBGL process is also boring and lengthy, which can result in poor learning attitudes [14].

Prensky (2001) defined the generation born after 1975 as the "G generation," with G standing for GAME, and advocated for learning through gaming. Such learning takes on the fun characteristics of games, thereby enhancing learning motivation and reducing the boredom of conventional classrooms. Therefore, learning through digital games is also known as joyful e-learning. Game-based e-learning is likely to become the primary learning method of the 21st century [17]. Gee (2003) also echoed Prensky's view that digital game learning has already changed the conventional passive acceptance of learning, with learners now becoming actively involved in the learning process [11]. The effect of digital games on students is evident. Students not only play games but also discuss, research, and imagine on the basis of these games [6]. Introducing digital games into courses causes the intrinsic motivation of learners to manifest as learning behavior, inspiring learners to devote more time and energy to learning [16]. Garris, Ahlers, and Driskell (2002) proposed a game learning model that included three stages—input, process, and outcome (Figure 1). During the game loop, because the performance of learners gradually improves, the pleasure gained from the game enhances learners' willingness to continue learning, creating a flow experience that results in a high degree of excitement and fulfillment [10]. Flow experiences occur when learners are inclined toward certain activities, becoming immersed in the activity to such an extent that they are not aware of the passage of time [5]. Numerous studies have indicated that digital game learning leads to more favorable learning outcomes and greater learning motivation than conventional learning does [1, 12, 15, 21, 23].

The essence of gamification is similar to that of game-based learning but differs in implementation. Designs that implement gamification emphasize the encouragement of users' motivation. They implement interesting and engaging elements that are derived from games and can be applied in real world activities. This process of application is referred to as "human-focused design," which comprises knowledge regarding optimized motivation, feeling, and participation [4]. Widely used gamification elements include points, badges, and leaderboards. In conventional teaching environments or systems, these elements are sufficient to motivate learners to compete with each other [20]. PeerWise is an after-school online learning system that enables students to design their own tests. In the system, badges are used to successfully increase student participation. The addition of gamification has been shown to increase students' enthusiasm for and use of the system [7]. Gamification applies game design items and features to nongaming content, and it can be employed in numerous educational environments, such as in providing learning outcomes (points, virtual coins, and ranks) to social interactions [8]. Gamification effectively increases learning intensity and enhances learning outcomes [13, 18]. Thus, it is an appropriate solution for increasing students' engagement and immersion in learning [9, 22].

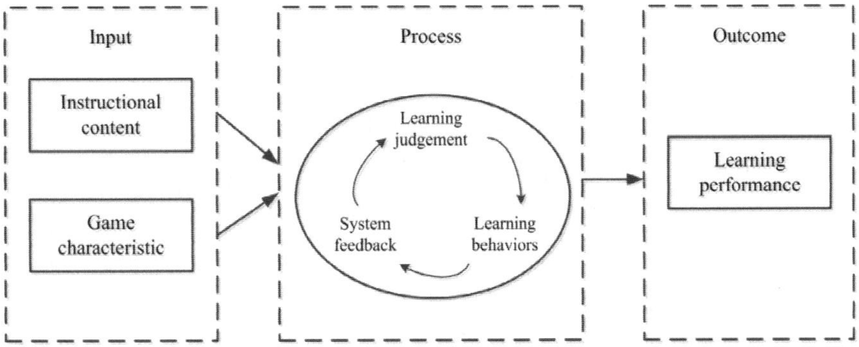

Figure 1. Gamification learning model (Garris, Ahlers, & Driskell, 2002).

Octalysis is a human-oriented gamification design framework [4]. As shown in Figure 2, the framework has an octagon shape, corresponding to eight core drives on each side. Octalysis arranges eight driving elements that generate learners' motivations, namely epic meaning and calling, development and accomplishment, empowerment of creativity and feedback, ownership and possession, social influence and relatedness, scarcity and impatience, unpredictability and curiosity, and loss and avoidance. The structure is often used to optimize learner perception, motivation, and engagement (participation).

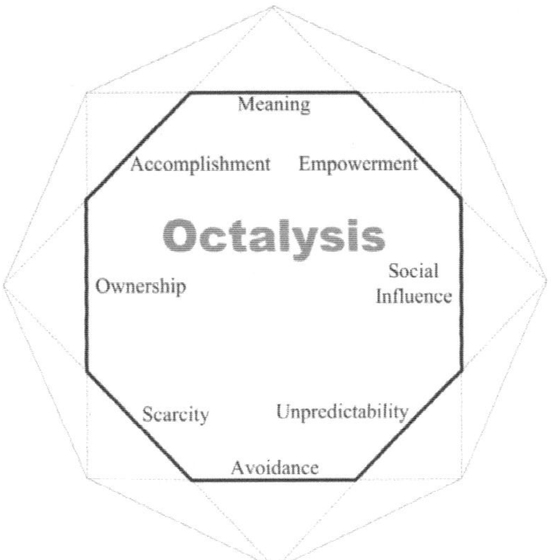

Figure 2. Octalysis gamification framework (Chou, 2015).

Training-based digital learning systems are designed for cognitive learning and can effectively improve learners' cognitive ability. However, the dullness of the learning process can often reduce learners' learning motivation. Moreover, the lack of guidance from teachers in asynchronous learning environments often leads learners to memorize cognitive content mechanically.

In this study, a novel gamification-based digital learning system was designed that employed asynchronous training. The CBGL algorithm was embedded in the system to provide personalized guided learning, enabling the system to take over the guiding role of teachers and providing students with an asynchronous digital learning environment. A gamification mechanism designed using Octalysis driving elements solved problems related to the dullness and length of the learning process, thereby increasing learners' motivation and engagement and generating a positive flow experience.

2. System design

2.1. System model

The system model is illustrated in Figure 3. The model had three primary parts, namely the user, an asynchronous HTML5-related gamified CBGL system, and a server for the database and web pages.

The system incorporated gamification core drives, the CBGL algorithm, and an HTML5-related cognitive e-course in an asynchronous training-based digital learning system, providing personalized guided learning and Octalysis gamification mechanisms.

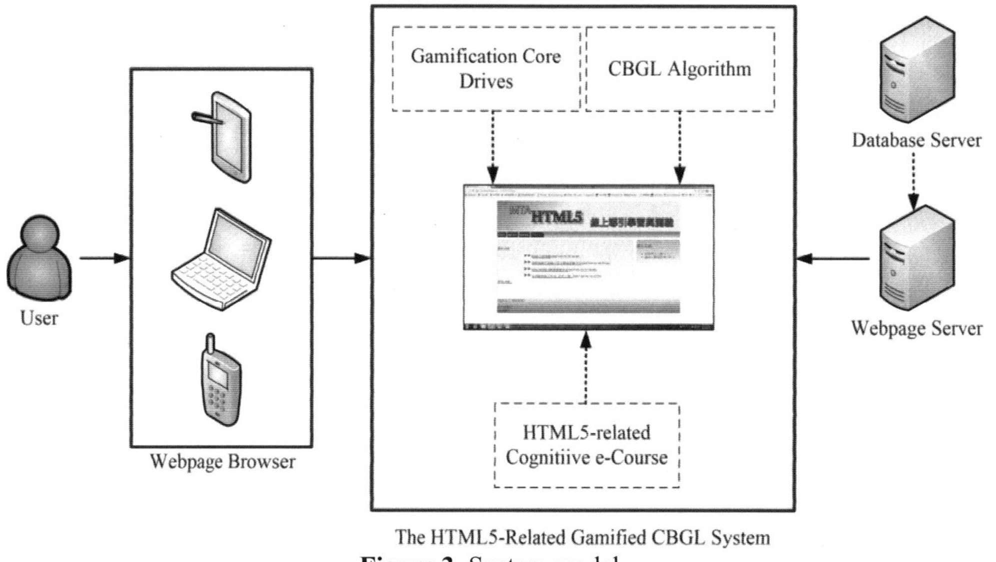

Figure 3. System model.

2.2. System structure

The system structure is shown in Figure 4. The system was divided into learner and administrator sections. The functions required by learners were explicitly established in the learner section. The administrator section provided researchers with functions for administration and study. The system website can be found at http://lcs.dk.ntc.edu.tw/~lcs/mtahtml5etask/index.php.

The features of the system were as follows:

1. An incorporated badge leaderboard.
2. Guided learning in the learner section that incorporated CBGL adaptive guidance and the gamification element of badges.
3. Octalysis development, namely a sense of achievement, gamification driving elements, and adaptive guidance functions.

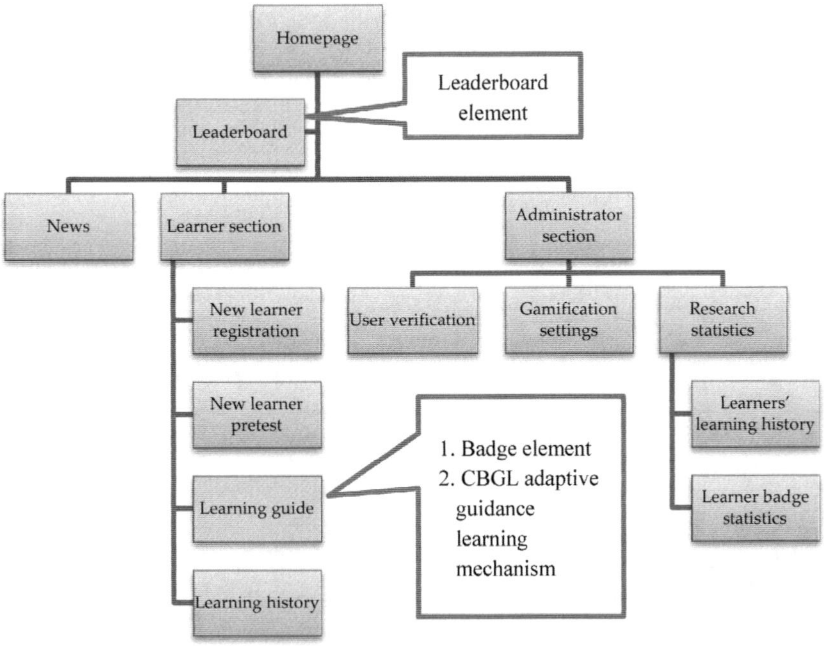

Figure 4. System architecture.

2.3. Design of the personalized cognitive learning system

Online test systems are often used to verify learning effects by using question banks as learning material. Programming is a standard course in information technology departments, but it is also a common subject in many other subject areas. Numerous programming languages exist, but for website development, HTML5 is the newest front-end programming language. Therefore, this study employed a cognitive question bank based on HTML5 for learning material and the CBGL algorithm to separate content into five learning units that included 257 questions. The system then constructed the e-course interface based on the included question bank.

If cognitive learning material is contained within a question bank, which is a fixed range and separated into five units, the CBGL algorithm can be used to generate personalized learning paths according to the different cognitive abilities of its users, with the system guiding users to study in stages [14]. Therefore, the CBGL algorithm was employed as the basis of the current study's system design. Every learner of the system underwent a pretest of their HTML5 cognitive abilities before learning, and the system generated a personalized learning path according to the pretest results.

Hsu and Li (2015) proposed the algorithmic process of a CBGL system (Figure 5) [14]. Upon completing registration, their system guides learners directly to the pretest stage. The system displays questions randomly from the question bank, and learners proceed through the test, with the result of every unit recorded in the database of the system. The obtained test result is then computed and generated to the learning path using Equation (1), which then proceeds to the drill and practice learning process. Learners' cognitive abilities must reach the required level to proceed to the next unit; otherwise, they must practice the unit until they attain the necessary standard.

$$Personal\ learning\ path ::= Bubble\ sort\ rank(Score(i)\,|_{1 \le i \le N}) \tag{1}$$

2.4. Design of gamification mechanisms

The literature indicates that the Octalysis framework, specifically the core driver of development and accomplishment, is suitable for most problem-based learning systems. Points, badges, and leaderboards have been used to raise users' engagement and motivation [4]. Moreover, these elements have been adopted extensively in gamification designs [20]. PeerWise, a famous online after-school learning system, also successfully used badges to improve students' engagement in the learning process [7].

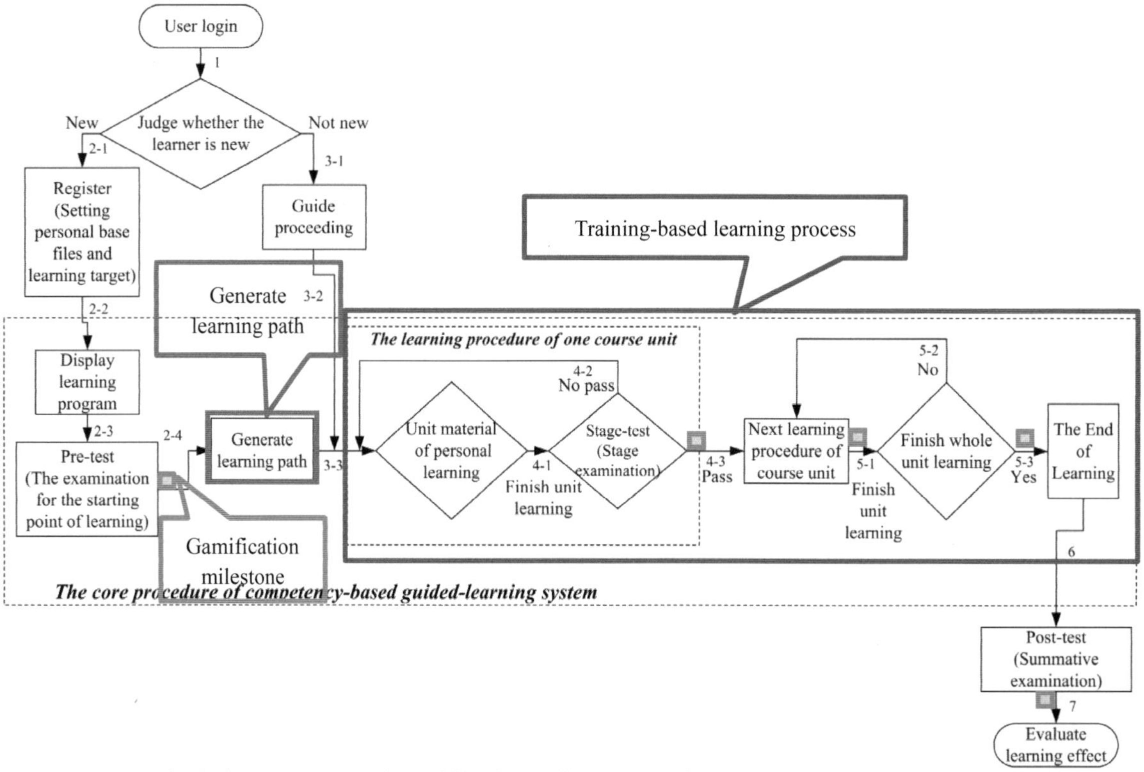

Figure 5. Calculation process and gamification milestone settings of the CBGL system (Hsu & Li, 2015).

This study focused on badges and leaderboards in Octalysis development under the core driver of development and accomplishment as the system gamification mechanism. Badges and leaderboards were made for each milestone of the adaptive CBGL path and, on the basis of the law of effect from behavior theory, were used to encourage users to complete various learning stages. The milestones were marked at units 2-4, 4-3, 5-1, 5-3, and 7, as shown in Figure 5. These milestones spanned the pretest process, learning process, and final summary test to prevent users from becoming bored. Assigning badges to each milestone was designed to keep users motivated by providing them with a sense of achievement as they progressed.

The incentive mechanism of the badges in this study was designed as follows:

1. CBGL
   (A) Obtain one badge for completing the pretest and beginning the learning path.
   (B) Obtain one badge for passing one level in the pretest.
   (C) Obtain one badge for passing one learning unit and deduct one badge for failing.
   (D) The final summary test verified learners' overall learning outcomes; thus, the number of available badges was also higher than the other tests—five badges were obtained if users passed the posttest summary test and five were deducted if they failed.
2. HTML5 simulation test

After completing the CBGL process, the simulation review test was initiated. The review test was for poststudy self-improvement; thus, if users passed the review test, they were awarded with two badges; if they failed, only one badge was deducted.

The calculation process of the badge incentive mechanism is shown in Figure 6. The system first extracted the table of user behavior logs from the database and then processed the following calculations:

Procedure 1: Determine whether the pretest procedure was completed. If completed, one badge is obtained.

Procedure 2: Determine whether the user already achieved the five goals in the pretest. If achieved, one badge is awarded for each goal.

Procedure 3: Determine the number of levels that the user passed during the learning process. Award the same number of badges as the number of levels passed, and deduct the same number of badges as the number of levels failed.

Procedure 4: Determine if the user completed the final summary test to verify the user's learning outcome. If the user passed the test, award five badges. Deduct five badges if the final summary test was failed.

Procedure 5: If the review test was completed, this indicates the completion of the entire guided learning process, and the system initiates the full-range simulation test. This test is an additional learning function to encourage the user to continue learning. Therefore, award two badges if the simulation test was passed; however, only deduct one badge if it was failed.

Procedure 6: Calculate the number of badges.

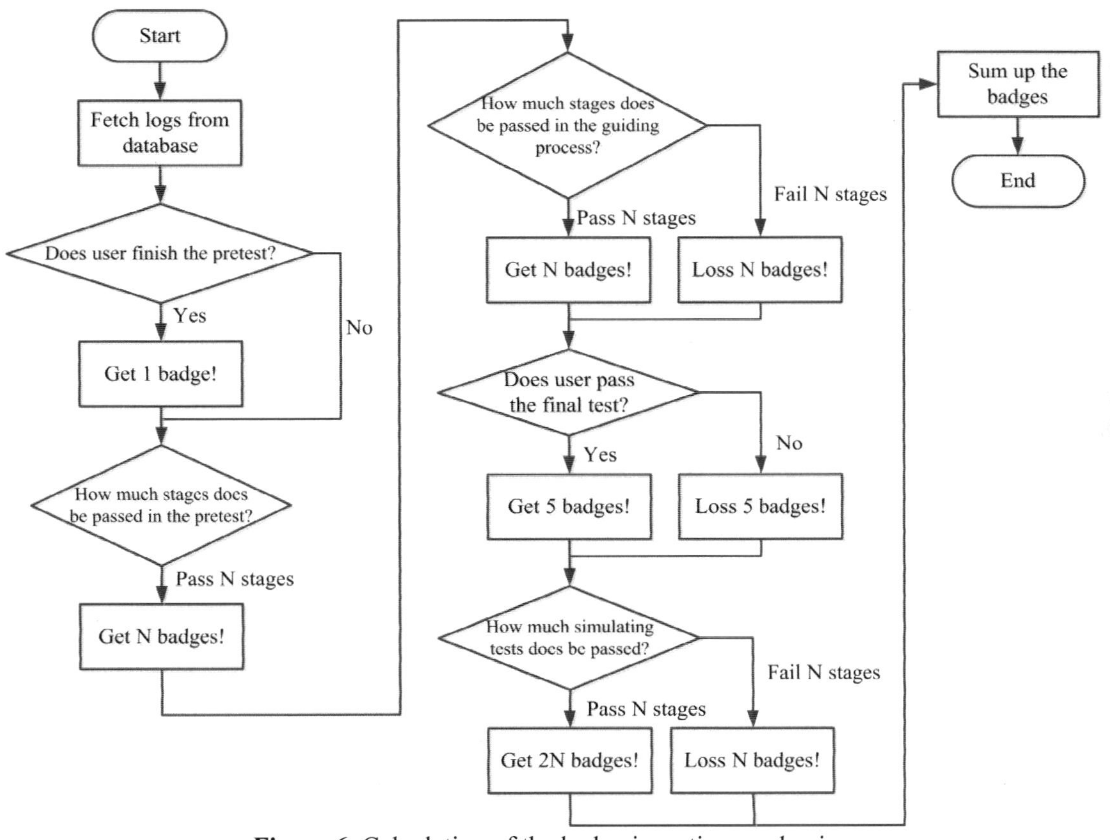

Figure 6. Calculation of the badge incentive mechanism.

The collection of badges was displayed in the user section of the system. A picture of the badge awarded is displayed in Figure 7. In addition to serving as evidence of student progression, the badges were used to positively encourage students to work hard. A badge leaderboard was also displayed on the system homepage to motivate users to progress together, as shown in Figure 8.

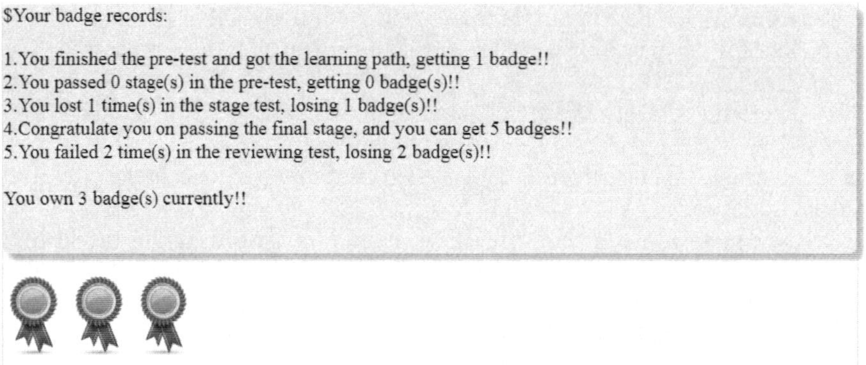

Figure 7. Realization of the badge incentive mechanism.

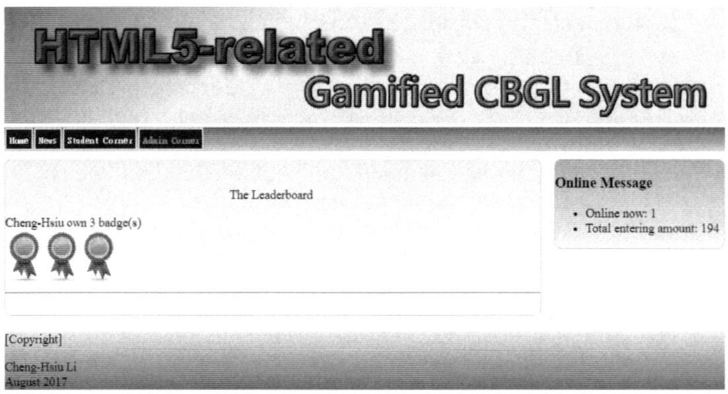

Figure 8. Image of the badge leaderboard.

3. Experiment results and discussion

3.1. Experimental scheme

1. Experimental period

The system experiment ran from September 20, 2017, to January 12, 2018.

2. Experimental participants

A total of 24 students who had taken the MTA HTML5 test at National Taitung Junior College in Taiwan were recruited for experimental testing.

3. Experimental course

The learning material in the experiment was the designed HTML5-related cognitive knowledge e-learning course.

4. Experimental design

(A) All of the participants proceeded through the learning process without teacher assistance, learning according to the system's adaptive guidance.

(B) Before the midterm examinations, the participants used the ungamified system for learning. Following the midterm examinations, they used the gamified system. The differences in the participants' completion of the CBGL process were subsequently explored using big data analysis.

3.2. Experimental analysis

1. Analysis of system satisfaction

A system satisfaction survey was conducted to determine learners' perceptions after using the system and whether the system fulfilled the research objectives. The adaptively guided learning system satisfaction questionnaire designed by Hsu and Li (2015) was chosen, and a 5-point Likert scale was used to examine the learners' satisfaction [14].

A total of 24 responses were collected for a response rate of 100%. The sixth item in the questionnaire was a reverse-scored question. After examining the reverse-scored question, one copy was eliminated, leaving 23 valid responses. The results for system satisfaction are listed in Table 1.

According to the results, the users felt the system caused them to willingly use it to actively learn. They also felt the learning process was engaging; thus, they were willing to patiently read through each question and complete all learning stages. Because the flow experience generated during the learning process was positive, the participants did not feel that the CBGL process was lengthy.

Table 1. Results from the system satisfaction survey.

Question	Mean	SD
I can use it to study positively.	3.91	0.733
The system can make me to learn.	3.87	0.548
I read every question patiently.	4.13	0.626
I feel the learning process is fun.	3.78	0.518
I finished all learning stages.	4.04	0.976
I feel the learning process is too lengthy.	2.96	0.825

Note: N=23.

2. Analysis of the completion of CBGL

Using big data analysis, users' completion of the CBGL process before and after using the gamification system was compared to reflect differences in the flow effect. The users' logs recorded into the database for big data analysis are illustrated in Figure 9. The results for comparison of the completion of CBGL are listed in Table 2.

The participants used the ungamified system for learning prior to the midterm examinations. According to the results of big data analysis, 14 students completed the learning sessions in the CBGL process, and the completion rate was 58.3%. After the midterm examinations, the participants switched to the gamification system. Subsequently, 19 students completed the learning sessions, and the completion rate was 79.2%. Thus, the completion rate increased by 20.9% after the system incorporated gamification mechanisms. This indicated that the incorporation of gamification helped the participants overcome the boring and lengthy learning process of the ungamified CBGL system.

id	user	time	catelog
31	sj985830@gmail.com	2017-11-22 18:17:40	Reading Guided-learning Question40
32	sj985830@gmail.com	2017-11-22 18:18:18	Reading Guided-learning Question41
33	sj985830@gmail.com	2017-11-22 18:18:22	Reading Guided-learning Question42
34	sj985830@gmail.com	2017-11-22 18:18:26	Reading Guided-learning Question43
35	sj985830@gmail.com	2017-11-22 18:18:29	Reading Guided-learning Question44
36	sj985830@gmail.com	2017-11-22 18:19:04	Reading Guided-learning Question45
37	love94630ji@gmail.com	2017-11-22 18:19:06	Pre-test3 testing
38	sj985830@gmail.com	2017-11-22 18:19:09	Reading Guided-learning Question46
39	sj985830@gmail.com	2017-11-22 18:19:11	Reading Guided-learning Question47
40	sj985830@gmail.com	2017-11-22 18:19:17	Reading Guided-learning Question48
41	sj985830@gmail.com	2017-11-22 18:19:22	Reading Guided-learning Question49
42	sj985830@gmail.com	2017-11-22 18:19:30	Reading Guided-learning Question50
43	sj985830@gmail.com	2017-11-22 18:19:35	Reading Guided-learning Question51
44	sj985830@gmail.com	2017-11-22 18:19:36	Reading Guided-learning Question52
45	sj985830@gmail.com	2017-11-22 18:19:44	Reading Guided-learning Question53

Figure 9. Recording users' logs for big data analysis.

Table 2. Comparison of the completion of CBGL.

Student No.	Using the ungamified system	Using the gamified system
1	Incomplete	Incomplete
2	Complete	Complete
3	Complete	Complete
4	Incomplete	Complete
5	Complete	Complete
6	Complete	Complete
7	Incomplete	Incomplete
8	Incomplete	Complete
9	Incomplete	Complete
10	Incomplete	Incomplete
11	Complete	Complete
12	Incomplete	Incomplete
13	Complete	Complete
14	Complete	Complete
15	Complete	Complete
16	Complete	Complete
17	Incomplete	Complete
18	Complete	Complete
19	Incomplete	Complete
20	Complete	Complete
21	Incomplete	Incomplete
22	Complete	Complete
23	Complete	Complete
24	Complete	Complete

4. Conclusions

Cognitive learning is the most fundamental method for humans to acquire knowledge, and training-based digital learning systems are designed to promote such learning. However, the dullness and length of training-based digital learning processes mean that poor learning motivation often affects learning outcomes. Training-based digital learning is normally conducted in an asynchronous manner. In other words, teachers and students are often not online at the same times, and therefore students must rely entirely on system assistance. This study successfully proposed an asynchronous gamified CBGL system that helps resolve the aforementioned problems.

The designed system for HTML5-related cognitive courses used the CBGL algorithm to provide a personalized guided learning environment for asynchronous digital learning that employed the gamified mechanisms of the Octalysis framework's core drivers to overcome dull and lengthy learning processes. The results indicated that the system helped users generate a positive flow experience that substantially improved their completion rates of CBGL tasks.

In summary, this system is recommended for future teaching experiments to explore whether the flow effect generated by gamification, which increased learning completion rates, can also produce a positive effect on learning outcomes.

References

[1] Adelman H S 1989 *Journal of Learning Disabilities* **22** pp 234 − 44
[2] Bergman E M, Sieben J M, Smailbegovic I, de Bruin A B, Scherpbier A J and van der Vleuten C P 2013 *Anatomical sciences education* **6** pp 114 − 24
[3] Burke R L 1982 *CAI sourcebook* (NJ: Pretice-Hall, Englewood Cliffs Press)
[4] Chou Y K 2015 *Gamification & Behavioral Design* http://yukaichou.com/gamification-examples/octalysis-complete-gamification-framework/

[5] Csikszentmihalyi M 1975 *Beyond boredom and anxiety* (San Francisco: Josey-Bass Press)
[6] De Castell S and Jenson J 2003 *Journal of Curriculum Studies* **35** pp 649–666
[7] Denny P 2013 The effect of virtual achievements on student engagement *In Proceedings of the SIGCHI conference on human factors in computing systems* pp 763–772
[8] Deterding S, Dixon D, Khaled R and Nacke L 2011 From game design elements to gamefulness: Defining ''Gamification'' *In Proceedings of MindTrek*
[9] Dondlinger M J 2007 *Journal of applied educational technology* **4** pp 21–31
[10] Garris R, Ahlers R and Driskell J E 2002 *Simulation & Gaming* **33** pp 441–467
[11] Gee J P 2003 *What video games have to teach us about learning and literacy* (New York: Palgrave/Macmillan Press)
[12] Gunstone R F and White R T 1981 *Science Education* **65** pp 291–299
[13] Hamari J and Koivisto J 2015 *Computers in Human Behavior* **50** pp 333–347
[14] Hsu W C and Li C H 2015 *Interactive Learning Environments* **23** pp 106–125
[15] Laffey J, Espinosa L, Moore J and Lodree A 2003 *Journal of Research on Technology in Education* **35** pp 423–440
[16] Malone T W and Lepper M R 1987 *Aptitude, learning, and instruction* **3** pp 223–253
[17] Prensky M 2001 *Digital game-based learning* (New York: McGraw-Hill Press)
[18] Simões J, Redondo R D and Vilas A F 2013 *Computers in Human Behavior* **29** pp 345–353
[19] Skinner B F 1968 *The technology of teach* (New York: Appleton Century Crofts Press)
[20] Song D, Yu P and Xu H 2017 *EURASIA Journal of Mathematics Science and Technology Education* **13** pp 3723–3734
[21] Terrell S and Rendulic P 1996 *Journal of Research on Computing in Education* **26** pp 403–414
[22] Von Ahn L and Dabbish L 2008 *Communications of the ACM* **51** pp 58–67
[23] Wexler S, Corti K, Derryberry A, Quinn C and van Barneveld A 2008 *Immersive learning simulations* (Santa Rosa: The eLearning Guild Research Press)
[24] Wikipedia 2017 Educational Technology https://en.wikipedia.org/wiki/Educational_technology

IOP Conf. Series: Materials Science and Engineering **658** (2019) 012005 doi:10.1088/1757-899X/658/1/012005

Evaluation of side resistance of driven precast concrete piles

M C Marcos[1] and Y-J Chen[2]

[1]Adamson University, Civil Engineering Department, Manila, Philippines
[2]Chung Yuan Christian University, Department of Civil Engineering, Taiwan

E-mail: yjc@cycu.edu.tw

Abstract. This study critically evaluates the axial side resistance of driven precast concrete (PC) piles. A wide range of load test data are classified into drained and undrained databases and subsequently used in the investigation. Each database is further divided into: (1) compression and uplift loading and (2) round and square cross sections. Measured and predicted results are both applied to examine the representative analytical models, including the alpha (α), beta (β), and lambda (λ) methods. The statistical results show that the range of values of the empirical coefficient α is wide. The predicted results of the β method are underestimated in drained loading, but they are reasonable in undrained loading. The relationship between λ and pile depth is also developed. Based on the analyses, the relative merits of the three analytical models are established, and designs for analyzing the side resistance of driven PC piles are recommended.

1. Introduction

Driven precast concrete (PC) piles are widely used as deep foundations for high rise buildings, towers, highway structures and others. Side resistance is an important source of driven PC pile capacity under axial loading, especially when the pile depth is considerably large or for piles under the condition of uplift loading in which the tip resistance is negligibly small. Research on side resistance of deep foundations has been progressing over the years. O'Neill [1] summarized results of significant recent research on a few aspects of side resistance for driven and drilled shafts while Lutenegger [2] focused on the importance of Standard Penetration Test for estimating driven pile side resistance. Another research [3, 4] provided an extensive evaluation of side resistance for drilled shafts using representative analytical models.

Analytical methods can be specified into alpha (α), beta (β), or lambda (λ) methods. Table 1 lists the analysis models and the related variables for each model. The α method [5] is a conventional total stress analysis for driven piles in cohesive soils. The side resistance is related to the average soil undrained shear strength (su) by an empirical coefficient denoted as α, which is the adhesion factor. The original α [5] for concrete piles was based on empirical correlations of mean su over the foundation depth while Stas and Kulhawy [6] developed α-su correlation for drilled shafts shown in Figure 1. In addition, some researchers [3, 4] develop the α-su correlation for a standardized undrained shear strength value from consolidated-isotropically undrained triaxial compression (CIUC) for drilled shafts. Furthermore, several researchers [7-9] also demonstrated that α is complexly related to other soil parameters such as the mean effective overburden stress ($\bar{\sigma}_{vm}$), overconsolidation ratio (OCR), and effective stress friction angle ($\bar{\phi}$).

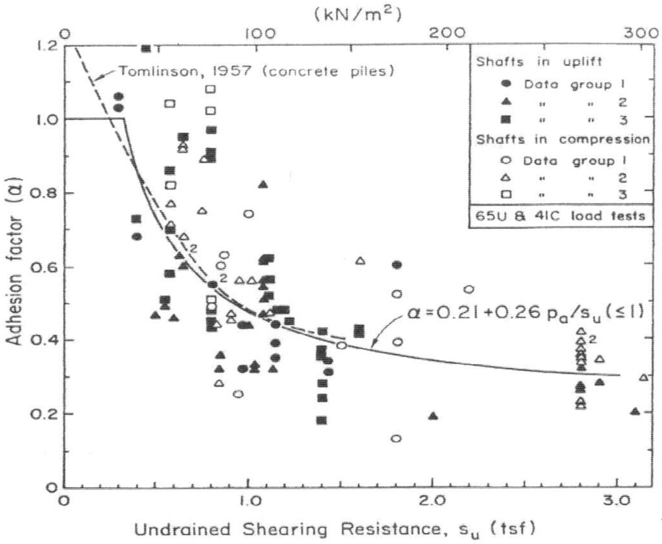

Figure 1. α - su Correlation [6]

Table 1. Analytical Models for Side Resistance Analysis

Method	**Analytical model**[a]		**Definition of factors**
α	$Q_s(\alpha) = p\sum_{n=1}^{N} \alpha_n s_{un} t_n$	(1)	α = empirical adhesion factor s_u = undrained shear strength
β	$Q_s(\beta) = p\left(\frac{K}{K_o}\right)\sum_{n=1}^{N} \bar{\sigma}_{vn} K_{on} \tan\left[\bar{\phi}_n \cdot \frac{\delta}{\bar{\phi}}\right] t_n$	(2)	K = coefficient of horizontal soil stress K_o = in-situ K $\bar{\sigma}_v$ = vertical effective stress $\bar{\phi}$ = effective stress friction angle δ = soil-shaft interface friction angle β = K tan δ
λ	$Q_s(\lambda) = \lambda p(\bar{\sigma}_{vm} + 2s_u)D$	(3)	λ = empirical factor

The β method [10] is an effective stress analysis which considers frictional resistance of the soil-shaft interface. In this method, side resistance is a function of the horizontal effective stress ($\bar{\sigma}_{ho}$), effective stress friction angle (δ) for the soil-shaft interface, and pile geometry. Previous study [11] examined available load test data and presented that the stress factor (K/K_o) is dependent on the construction method and its influence on in-situ stress. They suggested that for small displacement piles, K/K_o is in the range of 0.75 to 1.25, whereas for large displacement piles, K/K_o is in the range of 1.0 to 2.0. They also suggested that the ratio of interface friction angle (δ) to soil friction angle ($\bar{\phi}$) is in the range of 0.8 to 1.0 for precast piles.

Lastly, the λ method [12] is a combination of total and effective stress analyses that can be used for cohesive soils. In this method, side resistance is related to s_u and $\bar{\sigma}_{vm}$ by an empirical factor λ. The original λ was developed based on a database of driven pipe pile data and is a function of the total depth of the pile.

Although numerous methods have been critically evaluated for side resistance analysis of driven piles, more consistent approaches have been developed for assessing soil design parameters [13] that warrant a complete re-assessment of side resistance behavior of driven PC piles. In addition, more updated load test data exist nowadays that can be utilized for verifying estimated results. Furthermore, the effects of shape and installation methods on side resistance are of equal importance to examine. A broad database is utilized to assess the relative merits and suitability of each analytical model. Results are compared statistically and graphically. Subsequently, specific design recommendations are given for the use of these models to driven PC pile side resistance design.

2. Database of load tests

The load test data were collected from geotechnical literature and load test reports. The database developed in this study consists of 234 field axial load test conducted in 99 sites. Among these data, 72 sites with 154 load tests are loaded in compression while 27 sites with 80 load tests are loaded in uplift. The case histories cover a range of soil profiles and pile shapes. These load tests were conducted around the world at different points in time. The soil profile is categorized herein as drained or undrained, based on the predominant soil condition along the pile depth. The piles are round or square in cross-section. The load tests are divided into two groups, based on the loading condition, while each group is further subdivided using the two profile types and two cross-sections, with a total of eight categories.

For group 1 in compression: (1) drained compression with round piles (DCR) has 10 sites with 37 tests; (2) drained compression with square piles (DCS) has 23 sites with 44 tests; (3) undrained compression with round piles (UCR) has 18 sites with 37 tests; and (4) undrained compression with square piles (UCS) has 21 sites with 36 tests. For group 2 in uplift: (5) drained uplift with round piles (DUR) has 7 sites with 27 tests; (6) drained uplift with square piles (DUS) has 10 sites with 31 tests; (7) undrained uplift with round piles (UUR) has 3 sites with 11 tests; and (8) undrained uplift with square piles (UUS) has 7 sites with 11 tests. All of the load tests have almost complete geological data and load-displacement curves, and all were conducted on straight-sided, driven PC piles. Hence, these field data should reflect common field situations and can be representative for subsequent applications in engineering practice.

The basic information for the DCR, DCS, UCR, and UCS tests are given in Tables 2 to 5, respectively, whereas the information for the DUR, DUS, UUR, and UUS tests are given in Tables 6 to 9, respectively. The values of α, β_m, and λ were back-calculated using Eqs. 1 to 3 and the field load test results. The L_1-L_2 method [14, 15] which is a graphical construction method was adopted to measure the axial compression capacity. The method presents reasonable results for pile design based on previous studies [16-18].

Table 2. Basic Information and Side Resistance Analysis Result for Drained Compression Round Section (DCR) Tests

Site & Pile no.	Test site/Soil description along pile depth	Depth, D (m)	Dia., B (m)	K_o [a]	β_p [b]	Q_{scp} [c] (kN)	Q_{scm} [d] (kN)	Q_{scm}/Q_{scp}	β_m [e]	β_m/β_p	$(K/K_o)_m$ [f]
DCR_1	Arkansas; fine & silty sand	13.7	0.41	0.431	0.261	356	1292	3.63	0.947	3.63	3.17
DCR_2-1	Dramen, Norway; medium to coarse sand	8.0	0.28	0.670	0.304	86	203	2.36	0.718	2.36	2.09
DCR_2-2		16.0	0.28	0.544	0.247	257	347	1.35	0.333	1.35	1.20
DCR_2-3		7.5	0.28	0.677	0.308	78	155	1.98	0.610	1.98	1.76
DCR_2-4		11.5	0.28	0.583	0.265	160	250	1.57	0.415	1.57	1.39
DCR_2-5		15.5	0.28	0.544	0.247	242	355	1.47	0.362	1.47	1.30
DCR_2-6		19.5	0.28	0.513	0.253	380	516	1.36	0.344	1.36	1.20
DCR_2-7		23.5	0.28	0.472	0.258	551	734	1.33	0.344	1.33	1.17
DCR_3	Spain; fine silty sand	18.0	0.91	0.420	0.261	1431	2260	1.58	0.413	1.58	1.38
DCR_4-1	Miliao, Taiwan; silty sand	38.0	0.60	0.409	0.262	3007	5665	1.88	0.493	1.88	1.65
DCR_4-2		27.0	0.50	0.454	0.260	1316	2504	1.90	0.494	1.90	1.67
DCR_4-3		27.0	0.60	0.454	0.260	1579	4175	2.64	0.687	2.64	2.32
DCR_4-4		27.0	0.60	0.454	0.260	1579	4559	2.89	0.750	2.89	2.54
DCR_4-5		25.0	0.60	0.454	0.260	1370	3934	2.87	0.746	2.87	2.52
DCR_4-6		25.0	0.60	0.454	0.260	1370	4225	3.08	0.801	3.08	2.71
DCR_4-7		22.0	0.50	0.454	0.260	904	1926	2.13	0.554	2.13	1.87
DCR_4-8		30.0	0.50	0.454	0.260	1599	3864	2.42	0.628	2.42	2.12
DCR_5-1	Kaohsiung,	16.0	0.40	0.517	0.285	360	364	1.01	0.255	0.89	0.89

DCR_5-2	Taiwan; clayey silty sand	20.0	0.40	0.517	0.252	548	525	0.96	0.242	0.96	0.85
DCR_5-3		30.0	0.60	0.503	0.254	1778	1749	0.98	0.250	0.98	0.87
DCR_5-4		27.0	0.60	0.503	0.254	1461	1387	0.95	0.242	0.95	0.84
DCR_5-5		22.0	0.30	0.503	0.254	497	1726	3.48	0.884	3.48	3.07
DCR_5-6		22.0	0.40	0.503	0.254	662	1836	2.77	0.705	2.77	2.45
DCR_5-7		22.0	0.50	0.503	0.254	828	2588	3.13	0.795	3.13	2.76
DCR_5-8		34.0	0.60	0.503	0.254	2245	4320	1.92	0.489	1.92	1.70
DCR_5-9		34.0	0.60	0.503	0.254	2245	3309	1.47	0.375	1.47	1.30
DCR_6-1	Changbin, Taiwan; silty sand	25.0	0.60	0.431	0.261	1322	4007	3.03	0.792	3.03	2.66
DCR_6-2		23.0	0.50	0.431	0.261	943	3130	3.32	0.867	3.32	2.91
DCR_6-3		23.0	0.50	0.431	0.261	943	2983	3.16	0.826	3.16	2.77
DCR_7	N. Carolina; sandy/clayey silts	16.5	0.31	0.476	0.258	547	1516	2.77	0.715	2.77	2.44
DCR_8	NR; loose silt and sand	40.0	0.30	0.517	0.252	3028	2549	0.84	0.212	0.84	0.74
DCR_9	College Station, Texas; sand	10.4	0.91	0.454	0.260	735	2145	2.92	0.759	2.92	2.56
DCR_10-1	Chiayi, Taiwan; silty sand	27.0	0.60	0.454	0.260	1786	3917	2.19	0.570	2.19	1.93
DCR_10-2		10.0	0.50	0.505	0.264	230	737	3.20	0.843	3.20	2.82
DCR_10-3		30.0	0.50	0.454	0.260	1802	4734	2.63	0.683	2.63	2.31
DCR_10-4		20.0	0.50	0.466	0.259	825	2277	2.76	0.714	2.76	2.43
DCR_10-5		10.0	0.50	0.505	0.264	230	737	3.20	0.843	3.20	2.82

[a] K_o = in-situ K; [b] β_p = predicted beta value; [c] Q_{scp} = predicted compression side resistance; [d] Q_{scm} = measured compression side resistance; [e] β_m = measured beta value; [f] $(K/K_o)_m$ = back-calculated stress factor

Table 3. Basic Information and Side Resistance Analysis Result for Drained Compression Square Section (DCS) Tests

Site & Pile no.	Test site/Soil description along pile depth	Depth, D (m)	Dia., B (m)	K_o [a]	β_p [b]	Q_{scp} [c] (kN)	Q_{scm} [d] (kN)	Q_{scm}/Q_{scp}	β_m [e]	β_m/β_p	$(K/K_o)_m$ [f]
DCS_1	Florida; sand	18.8	0.762	0.389	0.261	1414	4123	2.92	0.762	2.92	2.54
DCS_2	Sweden; sand	12.8	0.235	0.727	0.398	300	328	1.09	0.434	1.09	0.96
DCS_3	Portugal; clayey sand	6.0	0.350	0.815	0.452	201	636	3.17	1.431	3.17	2.78
DCS_4-1	Kuwait; silty sand	12.1	0.300	0.420	0.261	283	298	1.05	0.275	1.05	0.92
DCS_4-2		9.3	0.300	0.409	0.262	176	181	1.03	0.269	1.03	0.90
DCS_5	Kuwait; sand	8.7	0.300	0.389	0.261	154	393	2.55	0.667	2.55	2.22
DCS_6	USA; fine sand	27.0	0.355	0.454	0.260	971	3119	3.21	0.834	3.21	2.82
DCS_7	Georgia; coarse sand	15.2	0.406	0.459	0.301	680	2186	3.21	0.968	3.21	2.80
DCS_8-1	Tidewater, Virginia; silty sand	21.0	0.360	0.420	0.261	917	1975	2.15	0.563	2.15	1.88
DCS_8-2		21.0	0.360	0.454	0.260	646	1639	2.54	0.659	2.54	2.23
DCS_8-3		26.0	0.360	0.472	0.270	828	3127	3.78	1.020	3.78	3.32
DCS_9-1	Iraq; sand	11.0	0.285	0.568	0.325	389	803	2.06	0.671	2.06	1.81
DCS_9-2		15.0	0.285	0.466	0.259	364	1229	3.37	0.874	3.37	2.97
DCS_10	China; sandy loam	25.0	0.450	0.454	0.260	1558	2363	1.52	0.394	1.52	1.33
DCS_11-1	Beijing; silty sand	10.0	0.300	0.437	0.261	490	1673	3.42	0.892	3.42	2.99
DCS_11-2		12.0	0.300	0.437	0.261	383	1055	2.75	0.718	2.75	2.41
DCS_12	USA; sand and silt	38.0	0.510	0.420	0.261	3053	7040	2.31	0.603	2.31	2.02
DCS_13-1	Ontario-site A; sand,silt and clay with some gravel	11.4	0.305	0.379	0.261	467	1313	2.81	0.734	2.81	2.44
DCS_13-2		11.2	0.305	0.379	0.261	453	1463	3.23	0.843	3.23	2.81
DCS_13-3		8.5	0.305	0.379	0.261	246	710	2.89	0.755	2.89	2.51
DCS_13-4		8.4	0.305	0.379	0.261	245	710	2.90	0.756	2.90	2.52
DCS_13-5		12.5	0.305	0.384	0.261	545	1463	2.68	0.700	2.68	2.33
DCS_13-6		15.1	0.305	0.346	0.258	749	1314	1.75	0.453	1.75	1.52
DCS_14-1	Ontario-site B; silty sand	34.8	0.305	0.431	0.261	1150	2435	2.12	0.553	2.12	1.85
DCS_14-2		16.5	0.305	0.442	0.261	300	820	2.73	0.712	2.73	2.39
DCS_15-1	Atlantic Coastal	13.0	0.254	0.518	0.270	168	422	2.52	0.680	2.52	2.22

DCS_15-2	Plain; loose to dense sand	10.0	0.356	0.456	0.276	173	329	1.91	0.526	1.91	1.67
DCS_15-3		16.5	0.356	0.465	0.259	521	1356	2.60	0.674	2.60	2.29
DCS_15-4		19.4	0.356	0.476	0.258	523	1835	3.51	0.904	3.51	3.09
DCS_15-5		19.2	0.457	0.478	0.258	661	1647	2.49	0.642	2.49	2.19
DCS_15-6		15.5	0.457	0.709	0.310	269	765	2.84	0.882	2.84	2.52
DCS_16	Netherland; coarse sand	9.3	0.250	0.591	0.298	209	491	2.35	0.700	2.35	2.07
DCS_17	Toronto; silty sand	14.7	0.305	0.466	0.259	571	776	1.36	0.352	1.36	1.19
DCS_18-1	Malaysia; clayey silty sand	46.8	0.400	0.478	0.258	2295	2768	1.21	0.311	1.21	1.06
DCS_18-2		48.5	0.400	0.454	0.260	3203	3951	1.23	0.321	1.23	1.08
DCS_19-1	Florida; loose to dense sand	13.4	0.600	0.442	0.261	1103	2785	2.52	0.658	2.52	2.21
DCS_19-2		10.5	0.600	0.475	0.264	840	2977	3.54	0.935	3.54	3.12
DCS_19-3		13.4	0.600	0.442	0.261	1103	2703	2.45	0.638	2.45	2.15
DCS_20	Virginia; clayey sand	9.4	0.508	0.478	0.258	600	1976	3.29	0.848	3.29	2.90
DCS_21-1	Netherlands; clayey and silty sand	30.6	0.380	0.442	0.261	1816	3183	1.75	0.457	1.75	1.54
DCS_21-2		30.3	0.380	0.442	0.261	1778	3362	1.89	0.493	1.89	1.66
DCS_21-3		30.7	0.380	0.442	0.261	1826	3778	2.07	0.623	2.39	2.10
DCS_22	Stockholm; loose sand	21.5	0.270	0.466	0.259	634	896	1.41	0.366	1.41	1.24
DCS_23	Virginia; fine sand	18.0	0.610	0.478	0.258	775	2506	3.23	0.833	3.23	2.85

[a] K_o = in-situ K; [b] β_p = predicted beta value; [c] Q_{scp} = predicted compression side resistance; [d] Q_{scm} = measured compression side resistance; [e] β_m = measured beta value; [f] $(K/K_o)_m$ = back-calculated stress factor

Table 4. Basic Information and Side Resistance Analysis Result for Undrained Compression Round Section (UCR) Tests

Site & Pile no.	Test site/Soil description along pile depth	Depth, D (m)	Dia, B (m)	OCR[a]	K_o[b]	β_p[c]	Q_{scp}[d] (kN)	Q_{scm}[e] (kN)	Q_{scm}/Q_{scp}	s_u (CIUC)[f]	α (CIUC)[g]	λ[h]
UCR_1	Boston; clay	45.5	0.41	2	0.664	0.379	4914	4325	0.88	158	0.47	0.14
UCR_2	Boston; clay	41.8	0.31	2	0.721	0.353	2826	1921	0.68	85	0.56	0.13
UCR_3	Philippines; sand	57.0	0.41	1	0.470	0.258	3129	2039	0.65	88	0.41	0.09
UCR_4-1	Puerto Rico; silty clay with sand	21.65	0.30	6	1.192	0.730	1782	1854	1.04	183	0.50	0.19
UCR_4-2		19.8	0.30	8	1.392	0.914	1893	1042	0.55	199	0.28	0.11
UCR_4-3		22.9	0.30	8	1.372	0.964	2617	2275	0.87	271	0.39	0.17
UCR_5	Brazil; silty clay	40.0	0.42	1	0.470	0.258	5352	3136	0.59	198	0.30	0.08
UCR_6-1	LA; sandy silty clay	13.0	0.46	6	1.232	0.556	1165	1499	1.29	104	0.77	0.25
UCR_6-2		17.7	0.46	4	0.990	0.524	2121	2084	0.98	171	0.48	0.16
UCR_7	Texas; stiff clay	31.0	0.32	3	0.927	0.367	2341	2898	1.24	175	0.53	0.17
UCR_8	Brazil; clayey silt	14.0	0.18	1	0.531	0.250	214	224	1.04	63	0.58	0.12
UCR_9-1	China; muck clay	38.2	0.40	1	0.500	0.255	2463	2550	1.04	95	0.56	0.14
UCR_9-2	China; sandy clay	24.8	0.40	5	1.103	0.606	2562	2613	1.02	119	0.70	0.22
UCR_9-3		19.6	0.30	7	1.318	0.725	1460	1500	1.03	113	0.72	0.24
UCR_10	Canada; silty clay	36.0	0.36	3	0.852	0.454	2510	3430	1.37	116	0.73	0.23
UCR_11	U.S.; clay	8.2	1.37	7	1.323	0.674	1753	2872	1.64	106	0.77	0.36
UCR_12-1	Malaysia; marine clay	35.5	0.25	1	0.531	0.250	865	545	0.63	38	0.81	0.07
UCR_12-2		14.5	0.25	2	0.735	0.346	230	274	1.19	45	1.00	0.23
UCR_12-3		23.5	0.25	1	0.531	0.250	404	441	1.09	59	0.63	0.13
UCR_12-4		11.5	0.25	2	0.735	0.346	177	179	1.01	40	0.99	0.20
UCR_13-1	Malaysia; silty clay	14.5	0.30	6	1.215	0.668	1083	1310	1.21	114	0.84	0.28
UCR_13-2	Malaysia; sandy clay	14.2	0.35	5	1.103	0.606	1077	1589	1.48	105	0.41	0.35
UCR_13-3	Malaysia; sandy clay	18.7	0.25	3	0.828	0.472	1024	874	0.85	145	0.41	0.14
UCR_14	Malaysia; clayey silt	28.5	0.40	1	0.459	0.259	2531	2163	0.85	159	0.38	0.10
UCR_15-1	Malaysia; soft marine clay	43.0	1.00	2	0.679	0.373	7405	6850	0.93	150	0.34	0.11
UCR_15-2		57.5	1.00	2	0.604	0.397	11689	7909	0.68	150	0.29	0.09
UCR_15-3		33.8	1.00	3	0.828	0.472	5701	5730	1.01	105	0.51	0.17

UCR_16	Malaysia; silty clay	42.0	0.60	3	0.889	0.418	4087	5493	1.34	72	0.96	0.26
UCR_17-1	Bangkok; soft to stiff clay with sand	28.0	0.60	4	0.980	0.539	4861	4266	0.88	102	0.79	0.22
UCR_17-2		29.0	0.80	3	0.841	0.463	6582	8264	1.26	107	0.82	0.28
UCR_17-3		38.0	0.80	2	0.679	0.373	7329	6969	0.95	108	0.68	0.17
UCR_17-4		24.6	0.80	4	0.980	0.539	5974	5651	0.95	104	0.88	0.24
UCR_17-5		29.4	0.80	3	0.841	0.463	9072	9255	1.02	118	0.82	0.25
UCR_18-1	Singapore; clayey silt	9.4	0.84	24	2.716	1.836	1138	2273	2.00	252	0.36	0.24
UCR_18-2		16.2	0.84	24	2.673	1.705	3693	5211	1.41	226	0.44	0.31
UCR_18-3		17.7	0.91	24	2.762	2.429	7340	5473	0.75	228	0.28	0.17
UCR_18-4		14.6	0.84	24	2.656	1.659	2787	3911	1.40	192	0.53	0.29

[a] OCR = overconsolidation ratio; [b] K_o = in-situ K; [c] β_p = predicted beta value; [d] Q_{scp} = predicted compression side resistance; [e] Q_{scm} = measured compression side resistance; [f] s_u(CIUC)= undrained shear strength from CIUC test; [g] α(CIUC) = calculated alpha; [h] λ = calculated lambda

Table 5. Basic Information and Side Resistance Analysis Result for Undrained Compression Square Section (UCS) Tests

Site & Pile no.	Test site/Soil description along pile depth	Depth, D (m)	Dia. B (m)	OCR[a]	K_o[b]	β_p[c]	Q_{scp}[d] (kN)	Q_{scm}[e] (kN)	Q_{scm}/Q_{scp}	s_u(CIUC)[f]	α(CIUC)[g]	λ[h]
UCS_1-1	Singapore; marine clay	26.0	0.280	2	0.735	0.346	1426	2038	1.43	46	1.54	0.38
UCS_1-2		30.0	0.260	2	0.735	0.346	1668	1716	1.03	45	1.50	0.29
UCS_2-1	Malaysia; silt and clay	18.2	0.300	5	1.059	0.678	2357	1983	0.84	111	0.55	0.20
UCS_2-2		17.4	0.350	5	1.059	0.678	2513	3061	1.22	109	0.77	0.28
UCS_2-3		18.6	0.350	4	0.928	0.594	2517	2662	1.06	112	0.61	0.22
UCS_2-4		14.7	0.300	6	1.179	0.755	1712	2412	1.41	101	0.72	0.27
UCS_2-5		16.8	0.250	5	1.059	0.678	1673	1621	0.97	107	0.59	0.22
UCS_3	Mexico; clayey soil	15	0.300	1	0.531	0.250	423	441	1.04	49	0.64	0.16
UCS_4-1	Tianjin, China; clayey soil	9.0	0.400	5	1.076	0.654	412	554	1.34	60	0.82	0.30
UCS_4-2		21.0	0.400	7	1.294	0.821	2308	2613	1.13	120	0.83	0.38
UCS_5-1	Singapore; clay and silt	25.1	0.280	2	0.650	0.384	2629	2164	0.82	180	0.55	0.16
UCS_5-2		25.6	0.325	2	0.650	0.384	3174	2339	0.74	204	0.44	0.14
UCS_6	Ireland; clayey silt	6.0	0.250	2	0.693	0.367	74	52	0.70	33	0.33	0.19
UCS_7-1	California; lean to fat clay	21.0	0.355	6	1.215	0.668	1885	2457	1.30	88	0.90	0.39
UCS_7-2		15.0	0.355	8	1.413	0.806	1163	1572	1.35	87	0.41	0.41
UCS_8	Mexico; soft clay	15.0	0.300	1	0.531	0.250	435	403	0.93	50	0.74	0.14
UCS_9-1	Guiana; Demerara clay	12.8	0.305	5	1.129	0.531	345	280	0.81	39	0.76	0.19
UCS_9-2		12.8	0.305	5	1.129	0.531	345	280	0.81	39	0.76	0.19
UCS_10	Georgia; sandy clay	22.9	0.457	7	1.323	0.622	2436	3071	1.26	114	0.82	0.29
UCS_11-1	London; London clay	8.5	0.305	14.4	1.847	0.855	698	702	1.01	146	0.59	0.23
UCS_11-2		4.4	0.305	25.3	2.679	1.579	328	291	0.89	127	0.54	0.23
UCS_12-1	London; London clay	13.1	0.400	7.5	1.370	0.645	1438	984	0.68	155	0.39	0.14
UCS_12-2	London; soft alluvium	12.8	0.360	7.8	1.392	0.655	1255	1275	1.02	154	0.57	0.21
UCS_12-3		12.2	0.360	8.4	1.437	0.676	1178	1020	0.87	153	0.48	0.18
UCS_13-1	Australia; silty clay	39.0	0.350	1	0.378	0.200	2896	3292	1.14	116	0.66	0.15
UCS_13-2		42.0	0.275	1	0.485	0.257	3304	3806	1.15	118	0.89	0.20
UCS_14	Louisiana; clays and silts	25.0	0.356	2	0.690	0.368	3120	1817	0.58	193	0.34	0.10
UCS_15-1	Indonesia; silt-clay	5.5	0.250	15	1.907	0.922	223	224	1.00	42	1.23	0.40
UCS_15-2		11.5	0.250	3	0.874	0.434	512	410	0.80	56	0.81	0.21
UCS_16	India; soft to stiff clay	22.5	0.400	3.7	0.916	0.541	2888	1472	0.51	168	0.31	0.11
UCS_17	Carbondale; silty clay	6.1	0.305	15	2.320	1.800	320	631	1.98	68	0.70	0.59

UCS_18	India; soft to stiff clay	19.5	0.400	4	0.975	0.545	1340	1184	0.88	87	0.56	0.19
UCS_19	Malaysia; med to stiff clay	60.3	0.406	1	0.483	0.257	6827	6133	0.90	123	0.65	0.15
UCS_20-1	Belgium; boom clay	7.4	0.350	9	1.472	0.637	434	796	1.83	106	0.92	0.41
UCS_20-2		11.6	0.350	8	1.408	0.663	1110	1350	1.22	121	0.88	0.31
UCS_21	Argentina; clay	19.0	0.400	2	0.672	0.376	1936	1855	0.96	44	1.77	0.25

[a] OCR = overconsolidation ratio; [b] K_o = in-situ K; [c] β_p = predicted beta value; [d] Q_{scp} = predicted compression side resistance; [e] Q_{scm} = measured compression side resistance; [f] s_u(CIUC)= undrained shear strength from CIUC test; [g] α(CIUC) = calculated alpha; [h] λ = calculated lambda

Table 6. Basic Information and Side Resistance Analysis Result for Drained Uplift Round Section (DUR) Tests

Site & Pile no.	Test site/Soil description along pile depth	Depth, D (m)	Dia., B (m)	K_o [a]	β_p [b]	Q_{sup} [c] (kN)	Q_{sum} [d] (kN)	Q_{sum}/Q_{sup}	β_m [e]	β_m/β_p	$(K/K_o)_m$ [f]
DUR_1	Spain; silty sand	18.0	0.91	0.412	0.262	1174	1986	1.69	0.434	1.66	1.45
DUR_2-1	Miliao, Taiwan; silty sand	25.0	0.50	0.441	0.261	1141	1429	1.25	0.325	1.25	1.09
DUR_2-2		27.0	0.50	0.441	0.261	1315	1799	1.37	0.355	1.36	1.19
DUR_2-3		23.0	0.50	0.441	0.261	979	1106	1.13	0.293	1.12	0.99
DUR_2-4		23.0	0.40	0.441	0.261	784	518	0.66	0.172	0.66	0.58
DUR_2-5		12.0	0.50	0.494	0.292	410	984	2.40	0.699	2.39	2.10
DUR_2-6		23.0	0.50	0.441	0.261	979	1534	1.57	0.407	1.56	1.37
DUR_2-7		23.0	0.50	0.441	0.261	979	1434	1.46	0.380	1.46	1.28
DUR_2-8		25.0	0.50	0.441	0.261	1141	1841	1.61	0.419	1.61	1.41
DUR_2-9		12.0	0.50	0.494	0.292	410	1087	2.65	0.772	2.64	2.32
DUR_2-10		11.0	0.50	0.415	0.245	295	809	2.74	0.671	2.73	2.40
DUR_3	Taiwan; silty sand	20.0	0.60	0.531	0.250	844	1483	1.76	0.455	1.82	1.61
DUR_4-1	Dramen, Norway; uniform loose normally consolidated sand	8.0	0.28	0.710	0.307	78	79	1.01	0.326	1.06	0.94
DUR_4-2		16.0	0.28	0.562	0.243	243	257	1.06	0.270	1.11	0.99
DUR_4-3		23.0	0.28	0.440	0.242	469	274	0.58	0.143	0.59	0.52
DUR_5-1	Brazil; silty sand	41.0	0.90	0.359	0.189	2560	1652	0.75	0.124	0.76	0.67
DUR_5-2		41.0	0.90	0.359	0.189	2560	1662	0.75	0.125	0.77	0.68
DUR_6-1	Vietnam; clayey sand	20.4	0.60	0.532	0.249	1890	2171	1.15	0.297	1.19	1.06
DUR_6-2	Vietnam; clayey sand	20.4	0.60	0.523	0.251	1902	2421	1.27	0.331	1.32	1.16
DUR_6-3	Vietnam; clayey sand	20.4	0.60	0.574	0.240	1815	1231	0.68	0.172	0.72	0.64
DUR_7-1	Miliao, Taiwan; silty sand	23.0	0.50	0.441	0.261	981	1341	1.37	0.355	1.36	1.19
DUR_7-2		9.0	0.50	0.441	0.261	279	939	3.37	0.875	3.36	2.94
DUR_7-3		9.0	0.50	0.441	0.261	279	1006	3.61	0.937	3.59	3.15
DUR_7-4		14.0	0.50	0.441	0.261	670	1367	2.04	0.530	2.03	1.78
DUR_7-5		14.0	0.50	0.441	0.261	670	1485	2.22	0.576	2.21	1.94
DUR_7-6		11.0	0.50	0.441	0.261	498	1588	3.19	0.994	3.81	3.34
DUR_7-7		23.0	0.50	0.441	0.261	1856	1658	0.89	0.232	0.89	0.78

[a] K_o = in-situ K; [b] β_p = predicted beta value; [c] Q_{sup} = predicted uplift side resistance; [d] Q_{sum} = measured uplift side resistance = Q_{12}-W; [e] β_m = measured beta value; [f] $(K/K_o)_m$ = back-calculated stress factor

Table 7. Basic Information and Side Resistance Analysis Result for Drained Uplift Square Section (DUS) Tests

Site & Pile no.	Test site/Soil description along pile depth	Depth D (m)	Dia.B (m)	K_o [a]	β_p [b]	Q_{sup} [c] (kN)	Q_{sum} [d] (kN)	Q_{sum}/Q_{sup}	β_m [e]	β_m/β_p	$(K/K_o)_m$ [f]
DUS_1	Iraq; uniform	11.0	0.285	0.455	0.260	312	408	1.31	0.341	1.31	1.15

	sand										
DUS_2	Canada; silty sand	14.7	0.305	0.345	0.190	423	255	0.60	0.116	0.61	0.54
DUS_3	Singapore; sand	23.1	0.400	0.412	0.262	2190	2473	1.13	0.290	1.11	0.97
DUS_4-1	USA; calcareous silty sand and sandy clay	19.4	0.762	0.469	0.259	1392	3848	2.76	0.722	2.79	2.46
DUS_4-2		16.7	0.610	0.483	0.257	826	2565	3.11	0.811	3.16	2.78
DUS_4-3		20.1	0.762	0.457	0.260	1492	3843	2.58	0.671	2.59	2.27
DUS_4-4		20.4	0.762	0.455	0.260	1537	4032	2.62	0.684	2.63	2.31
DUS_4-5		15.8	0.610	0.489	0.256	739	2271	3.07	0.803	3.13	2.76
DUS_4-6		16.4	0.610	0.485	0.257	796	2438	3.06	0.800	3.11	2.74
DUS_4-7		17.7	0.762	0.478	0.258	1159	2374	2.05	0.535	2.08	1.83
DUS_4-8		13.8	0.610	0.561	0.286	632	1263	2.00	0.585	2.05	1.81
DUS_4-9		13.5	0.610	0.566	0.288	610	1043	1.71	0.505	1.75	1.55
DUS_4-10		14.1	0.762	0.500	0.255	735	1295	1.76	0.460	1.81	1.59
DUS_4-11		17.7	0.610	0.478	0.258	928	2757	2.97	0.776	3.01	2.65
DUS_5-1	Brazil; sandy soil	12.0	0.170	0.475	0.258	69	24	0.34	0.089	0.35	0.30
DUS_5-2		12.0	0.170	0.475	0.258	69	38	0.54	0.142	0.55	0.49
DUS_5-3		12.0	0.170	0.475	0.258	69	54	0.78	0.203	0.79	0.69
DUS_5-4		12.0	0.170	0.475	0.258	82	46	0.56	0.146	0.56	0.50
DUS_6-1	California; silty sands	15.24	0.356	0.541	0.322	285	900	3.16	1.013	3.14	2.76
DUS_6-2		15.1	0.356	0.543	0.323	282	869	3.09	0.993	3.07	2.69
DUS_6-3		17.7	0.356	0.505	0.301	371	908	2.45	0.733	2.44	2.14
DUS_6-4		17.4	0.356	0.509	0.303	360	746	2.07	0.625	2.06	1.81
DUS_7-1	California; soft to dense sands,	11.6	0.356	0.608	0.362	186	481	2.58	0.931	2.57	2.25
DUS_7-2		12.2	0.356	0.600	0.357	200	631	3.15	1.121	3.14	2.75
DUS_7-3		12.2	0.356	0.600	0.357	200	418	2.09	0.742	2.08	1.82
DUS_8-1	India; sandy silt	3.0	0.200	0.720	0.353	26	29	1.13	0.413	1.17	1.03
DUS_8-2		3.0	0.200	0.720	0.353	26	29	1.13	0.413	1.17	1.03
DUS_8-3		3.0	0.200	0.720	0.353	26	36	1.41	0.512	1.45	1.28
DUS_9-1	USA; clayey sand	23.5	0.356	0.502	0.255	638	632	0.99	0.258	1.01	0.90
DUS_9-2		16.2	0.457	0.529	0.250	462	313	0.68	0.176	0.70	0.62
DUS_10	USA; fine silty sand	17.0	0.305	0.485	0.257	280	377	1.35	0.352	1.37	1.21

[a] K_o = in-situ K; [b] β_p = predicted beta value; [c] Q_{sup} = predicted uplift side resistance; [d] Q_{sum} = measured uplift side resistance = Q_{t2}-W; [e] β_m = measured beta value; [f] $(K/K_o)_m$ = back-calculated stress factor

Table 8. Basic Information and Side Resistance Analysis Result for Undrained Uplift Round Section (UUR) Tests

Site & Pile no.	Test site/Soil description along pile depth	Depth, D (m)	Dia. B (m)	OCR[a]	K_o[b]	β_p[c]	Q_{sup}[d] (kN)	Q_{sum}[e] (kN)	Q_{sum}/Q_{sup}	s_u(CIUC)[f]	α(CIUC)[g]	λ[h]
UUR_1-1	Negeri Sembilan, Malaysia; clay	17.5	0.50	1	0.562	0.243	984	600	0.61	37	0.67	0.099
UUR_1-2		17.5	0.50	1	0.562	0.243	984	1120	1.14	27	1.62	0.184
UUR_1-3		7.5	0.50	2	0.741	0.342	245	360	1.47	27	1.24	0.271
UUR_2-1	Bangkok, Thailand; soft clay	20.0	0.40	1	0.368	0.173	350	365	1.04	39	0.43	0.092
UUR_2-2		20.0	0.40	1	0.368	0.173	350	215	0.61	39	0.48	0.054
UUR_3-1	China; silty clay	43.0	0.50	1	0.172	0.081	941	631	0.67	20	0.59	0.039
UUR_3-2		43.0	0.50	1	0.172	0.081	941	631	0.67	20	0.59	0.039
UUR_3-3		43.0	0.50	1	0.172	0.081	941	786	0.84	20	0.70	0.049
UUR_3-4		43.0	0.60	1	0.172	0.081	1130	897	0.79	20	0.70	0.046
UUR_3-5		43.0	0.60	1	0.172	0.081	1130	927	0.82	20	0.72	0.048
UUR_3-6		43.0	0.60	1	0.172	0.081	1130	933	0.83	20	0.72	0.048

[a] OCR = overconsolidation ratio; [b] K_o = in-situ K; [c] β_p = predicted beta value; [d] Q_{sup} = predicted uplift side resistance; [e] Q_{sum} = measured uplift side resistance = Q_{t2}-W; [f] s_u(CIUC)= undrained shear strength from CIUC test; [g] α(CIUC) = calculated alpha; [h] λ = calculated lambda

Table 9. Basic Information and Side Resistance Analysis Result for Undrained Uplift Square Section (UUS) Tests

Site & Pile no.	Test site/Soil description along pile depth	Depth, D (m)	Dia., B (m)	OCR[a]	K_o[b]	β_p[c]	Q_{sup}[d] (kN)	Q_{sum}[e] (kN)	Q_{sum}/Q_{sup}	s_u(CIUC)[f]	α(CIUC)[g]	λ[h]
UUS_1	Mexico; clayey soil	15.0	0.300	1	0.531	0.250	386	363	0.94	48	0.58	0.141
UUS_2-1	Canada-Site A; glacial clay	25.0	0.380	1	0.212	0.108	857	615	0.72	78	0.30	0.056
UUS_2-2		23.8	0.380	1	0.224	0.114	819	629	0.77	76	0.33	0.063
UUS_3	Canada-Site B; marine silt	47.2	0.380	1	0.100	0.047	1266	1090	0.86	50	0.44	0.034
UUS_4	Louisiana; clays and silt	25.0	0.356	2	0.234	0.124	833	477	0.57	182	0.19	0.031
UUS_5-1	Illinois; silty clay	6.4	0.305	12	1.744	0.923	265	381	1.44	75	0.56	0.356
UUS_5-2		6.4	0.305	10	1.588	0.841	241	303	1.26	75	0.56	0.283
UUS_5-3		6.4	0.305	10	1.588	0.841	241	191	0.79	75	0.45	0.178
UUS_6-1	Kinnegar N. Ireland; clayey silt	6.0	0.250	2	0.693	0.367	81	55	0.68	30	0.45	0.150
UUS_6-2		6.0	0.250	2	0.693	0.367	81	52	0.65	30	0.43	0.145
UUS_7	Singapore; clay & silt	11.6	0.320	2	0.693	0.367	616	597	0.97	121	0.44	0.173

[a] OCR = overconsolidation ratio; [b] K_o = in-situ K; [c] β_p = predicted beta value; [d] Q_{sup} = predicted uplift side resistance; [e] Q_{sum} = measured uplift side resistance = Q_{l2}-W; [f] s_u(CIUC)= undrained shear strength from CIUC test; [g] α(CIUC) = calculated alpha; [h] λ = calculated lambda

The load-displacement curve in Figure 2 can generally be simplified into three distinct regions: initial linear, curve transition, and final linear. Point L_1 (elastic limit) corresponds to the load (Q_{L1}) and butt displacement (ρ_{L1}) at the end of the initial linear region, while L_2 (failure threshold) corresponds to the load (Q_{L2}) and butt displacement (ρ_{L2}) at the initiation of the final linear region. Q_{L2} is defined as the "interpreted failure load or capacity" because beyond Q_{L2}, a small increase in load gives a significant increase in displacement. The interrelationships between L_2 and other interpretation criteria from the lower to higher bounds that was previously developed [17] are used to infer the required L_2 if the test data are insufficient or are terminated prematurely.

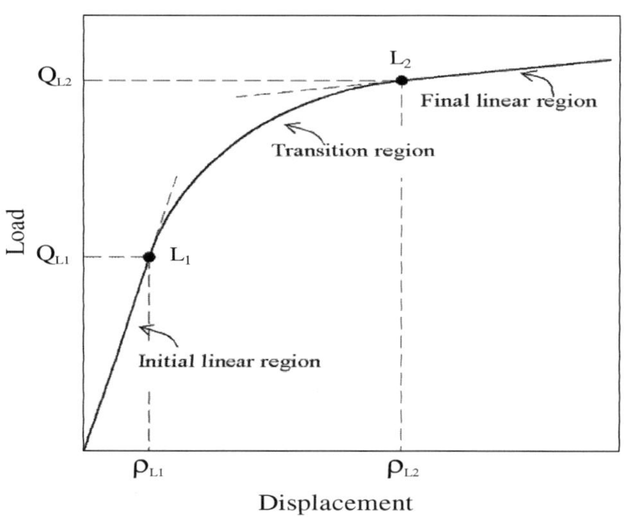

Figure 2. Regions of Axial Load-Displacement Curve [14]

The reference sources are too lengthy to include in this paper but are found elsewhere [19]. Each pile is identified by a site and pile numbers therein. For convenience, the ranges of foundation geometry, side resistance, and their statistics are summarized in Tables 10 and 11 for compression and uplift, respectively. As can be seen, the range of geometry is broad and the diameters for the four categories are roughly comparable.

Table 10. Range of Driven Pile Geometry for Compression Side Resistance Analysis

Data	Number of tests	Statistics	Pile geometry (m)		D/B	Side[b] resistance (KN)
			Depth, D	Diameter[a], B		
		Range	7.5-40.0	0.28-0.91	11.4-133.3	78-3028
DCR	37	Mean	22.1	0.48	48.6	1062
		COV	0.37	0.33	0.43	0.73
		Range	6.0-48.5	0.24-0.76	17.1-121.2	154-3203
DCS	44	Mean	17.9	0.38	49.3	819
		COV	0.56	0.30	0.52	0.89
		Range	8.2-57.5	0.18-1.37	6.0-142.0	177-11689
UCR	37	Mean	26.9	0.50	63.7	3355
		COV	0.48	0.57	0.55	0.82
		Range	4.4-60.3	0.25-0.46	14.5-152.7	74-6827
UCS	36	Mean	18.7	0.33	56.6	1647
		COV	0.61	0.17	0.61	0.81

[a] or width of square section; [b] from β method

3. Database of load tests

3.1. α method

3.1.1. αCIUC-su(CIUC)/pa correlations. The value of α for driven PC piles can be back-calculated from the field load test results as follows:

Table 11. Range of Driven Pile Geometry for Uplift Side Resistance Analysis

Data	Number of tests	Statistics	Pile geometry (m)		D/B	Side[b] resistance (KN)
			Depth, D	Diameter[a], B		
		Range	8.0 - 41.0	0.30 - 0.90	19.8 - 82.1	78 - 2560
DUR	27	Mean	19.8	0.50	39.7	1007
		COV	0.42	0.30	0.34	0.69
		Range	3.0 - 23.5	0.20 - 0.80	15.0 - 70.6	26 - 2190
DUS	31	Mean	14.5	0.40	38.3	561
		COV	0.35	0.48	0.47	0.95
		Range	7.5 - 43.0	0.40 - 0.60	15.0 - 86.0	81 - 1266
UUR	11	Mean	31.0	0.50	59.8	517
		COV	0.46	0.14	0.41	0.75
		Range	6.0 – 47.2	0.30 - 0.40	21.0 - 124.2	245 - 1130
UUS	11	Mean	16.3	0.30	47.3	830
		COV	0.80	0.15	0.68	0.41

[a] or width of square section; [b] from β method

$$\alpha = \frac{Q_s(L_2)}{pDs_u} \tag{1}$$

in which Q_s (L_2) = interpreted side resistance using L_2 method and s_u = mean undrained shear strength over the pile depth (D), and p = perimeter. To standardize the α-s_u relationship for driven piles, the unique test type of undrained shear strength from consolidated-isotropically undrained triaxial compression (CIUC) [3] was adopted. The CIUC was selected as reference test because it is quite common and of good quality test. The s_u values from all other tests were converted to "equivalent"

s_u(CIUC). The procedures to convert are based on previous study [20] for unconsolidated-undrained triaxial (UU) and unconfined compression (UC) tests.

Figure 3 illustrates the correlations between α and undrained shear strength for compression and uplift loading, with regression equations given as:

Compression:

$$\alpha_{(CIUC)} = 0.33 + 0.32 / [s_u(CIUC)/p_a] \quad (n = 73; SD = 0.20; r^2 = 0.54) \tag{2}$$

Uplift:

$$\alpha_{(CIUC)} = 0.22 + 0.14 / [s_u(CIUC)/p_a] \quad (n = 22; SD = 0.25; r^2 = 0.32) \tag{3}$$

All data:

$$\alpha_{(CIUC)} = 0.35 + 0.15 / [s_u(CIUC)/p_a] \quad (n = 95; SD = 0.28; r^2 = 0.42) \tag{4}$$

The s_u(CIUC) is normalized by p_a, which is the atmospheric stress (101.3 kN/m^2) in the same unit as s_u. As can be seen in Figure 3, the trends of round and square piles for compression loading are somewhat comparable. The same is true for uplift loading. Comparisons of compression and uplift α_{CIUC}-s_u(CIUC) correlations show some interesting points. First, the coefficient of determination (r^2) is larger for compression which may be due to the limited number of data used for uplift. Second, the compression data points are seen above uplift data points indicating that α_{CIUC} values are generally smaller for uplift for a range of s_u values. Third, Figure 3 demonstrates that α for small values of s_u(CIUC)/p_a produces steep regression lines for compression and uplift. Hence, the use of α_{CIUC}-s_u(CIUC) correlation for design may tend to be conservative and sensitive for relatively small s_u. Based on the available load test data, α for compression is in the range of 0.28 to 1.77, whereas for uplift, α is in the range of 0.19 to 1.62.

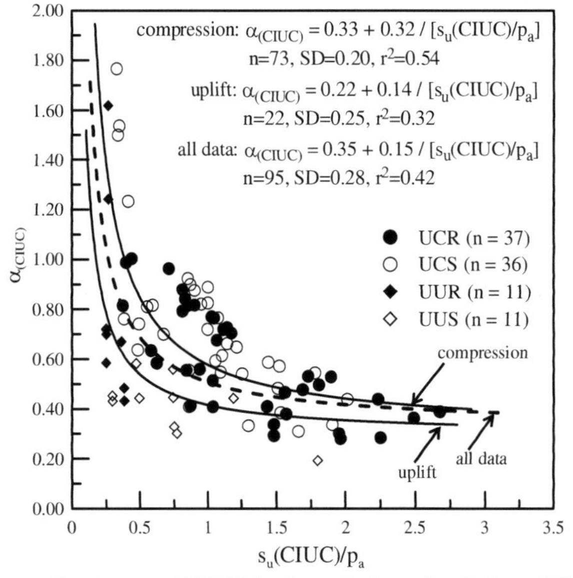

Figure 3. α_{CIUC} - s_u(CIUC)/p_a Correlations for Driven Piles

The α_{CIUC}-s_u(CIUC) correlation for drilled shafts [4] is compared to the α_{CIUC}-s_u(CIUC) correlation developed in this study. The α_{CIUC}-s_u(CIUC) correlation by the previous study [4] was developed from compression and uplift data. For comparison purposes, the compression and uplift α_{CIUC}-s_u(CIUC) correlations were combined. The comparison is presented in Figure 4 wherein some significant points are observed. First, the coefficient of determination (r^2) is higher in drilled shaft than in driven pile. This implies that consistent α value can be expected from drilled shaft than from driven pile. This behavior can be attributed to the better adhesion of soil-pile interface of drilled shaft which is the result of its installation procedure. The effect of hammer pile driving on the pore water pressure may indirectly affect the soil-pile adhesion resulting to more variable behavior of driven pile. Second, for smaller $s_u(CIUC)/p_a$ (< 0.50), the regression lines of drilled shaft and driven pile are converging. However, for larger values of $s_u(CIUC)/p_a$, drilled shaft tends to produce smaller values of α than driven pile. However, the difference is small. Third, the range of α for drilled shaft is smaller (with a maximum of < 1.0) than for driven pile (with a maximum of < 1.8).

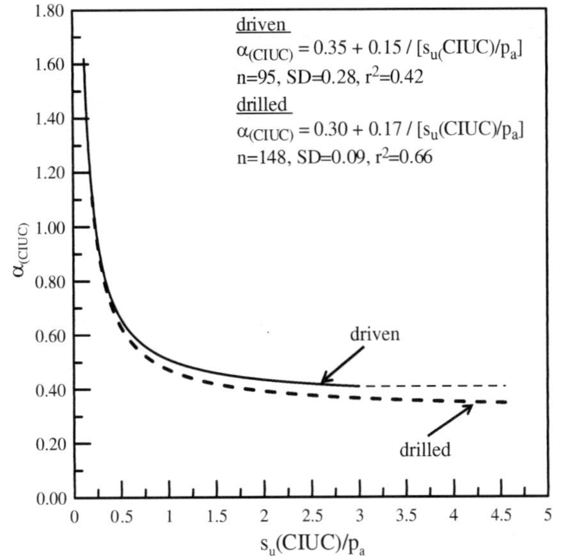

Figure 4. Comparison of Driven Pile and Drilled Shaft α_{CIUC} - $s_u(CIUC)/p_a$ Correlations

3.1.2. α_{CIUC} - $s_u(CIUC)/\bar{\sigma}_v$ - $\bar{\sigma}_v$ correlations. The undrained shear strength ratio correlation, α_{CIUC}-$s_u(CIUC)/\bar{\sigma}_v$ - $\bar{\sigma}_v$ was developed in which the overburden pressure was taken into account. The correlations were developed directly from field load test database and are presented in Figures. 5(a) to 5(c) for compression, uplift, and all data combined. These figures also include the statistical data for the individual curve. Result for compression loading in Figure 5(a) shows that the regression lines are closer when the undrained shear strength ratio (USR) is greater than 1.0. This could be due to a more scattered data for USR <1.0. Although the data points are few for small values of $\bar{\sigma}_v$, the regression lines appear to be stiffer for these values. This behavior is similar to the result in Figure 3 that the design may tend to be conservative for small values of $\bar{\sigma}_v$. Furthermore, α_{CIUC} is decreasing with increasing $s_u(CIUC)/\bar{\sigma}_v$ and $\bar{\sigma}_v$. For uplift, the data were subdivided into two ratios only due to limited number of data points and the correlation is shown in Figure 5(b). Roughly, similar behavior is observed as in compression that the trend of α_{CIUC} is decreasing with increasing $s_u(CIUC)/\bar{\sigma}_v$ and $\bar{\sigma}_v$. The compression and uplift data were combined as shown in Figure 5(c) which indicates a wide variation of data for USR < 0.50.

The α_{CIUC} value from α_{CIUC}-$s_u(CIUC)/\bar{\sigma}_v$ - $\bar{\sigma}_v$ correlations can precisely be distinguished than using the conventional α_{CIUC}-$s_u(CIUC)$ correlations. Therefore, the necessary α value can be reasonably selected. These correlations can be regarded as an alternative method of analysis for traditional α - s_u correlations or in verifying the required value of α for design.

3.2. β method
Approximate beta values can be predicted as follows:

$$\beta_p = K_o\,(K/K_o)\,\tan\,[\bar{\phi}\,\cdot\,\delta\,/\,\bar{\phi}\,] \tag{5}$$

Then, the average β_p over the pile depth was calculated by weighted average. The average $\bar{\phi}$ and K_o are shown in Tables 2 to 5 and Tables 6 to 9 for compression and uplift loading, respectively. In this study, $\delta / \bar{\phi}$ was taken as 0.90, and therefore all cases used $\delta / \bar{\phi}$ = 0.90 for this calculation. For K/K_o, a value of 1.0 was adopted which is a value within the range of small to large displacement piles. The detailed analysis results for the DCR, DCS, UCR, and UCS tests are given in Tables 2 to 5, respectively, whereas the results for the DUR, DUS, UUR, and UUS tests are given in Tables 6 to 9, respectively.

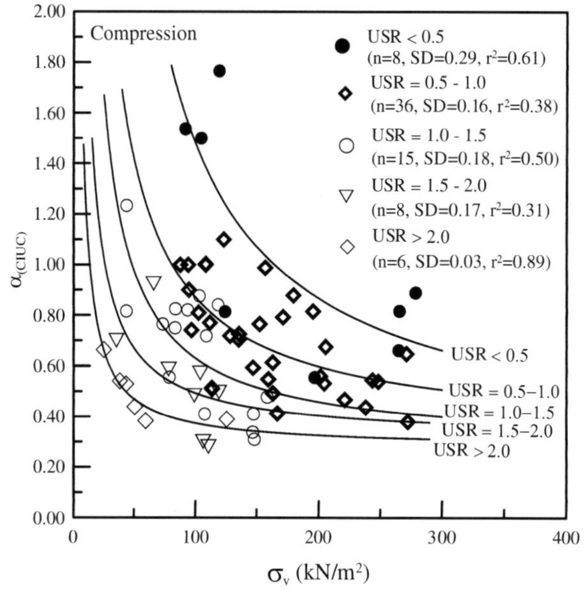

(a)

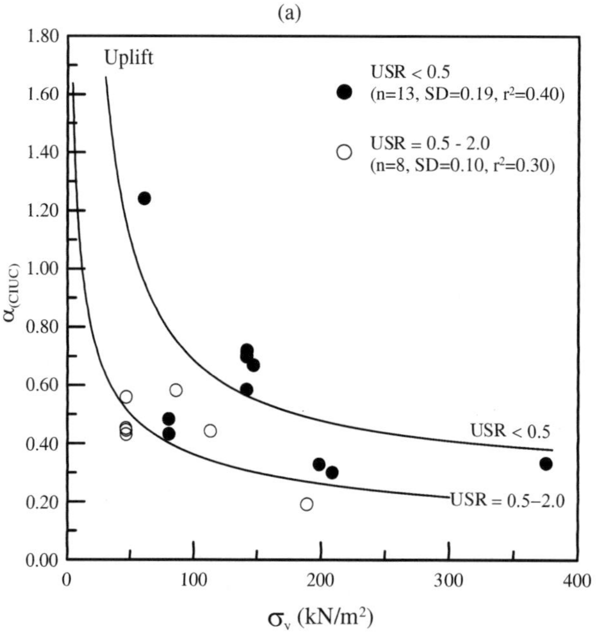

(b)

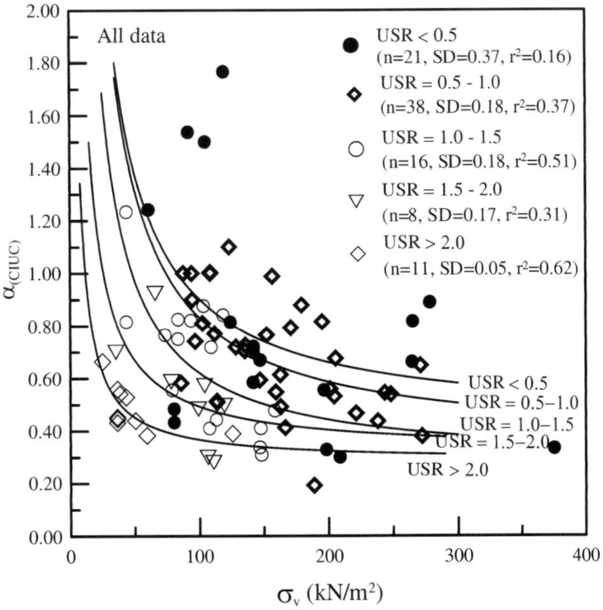

(c)

Figure 5. α_{CIUC} - $s_u(CIUC)/\bar{\sigma}_v$ - $\bar{\sigma}_v$ Correlations for (a) Compression Loading, (b) Uplift Loading, and (c) All data

3.2.1. Drained load tests. The result for the drained loading is summarized in Table 12 including the statistics. For group 1 (compression loading), the predicted side resistance (Q_{scp}) is compared to the measured side resistance (Q_{scm}). The mean side resistance ratios (Q_{scm}/Q_{scp}) are 2.25 and 2.44 for round and square cross section piles, respectively. The COVs for both results are more than 30%. The capacity ratios indicate an obvious underprediction of side resistance for both pile sections. For uplift, the predicted side resistance (Q_{sup}) likewise is compared to the measured side resistance (Q_{sum}) as shown in Table 12. The mean side resistance ratios (Q_{sum}/Q_{sup}) are 1.64 and 1.88, for round and square cross section piles, respectively. The COVs for both results are around 50%. As in compression, the capacity ratios indicate an obvious underprediction of side resistance for both pile sections.

Table 12. Statistics of Q_{sm}/Q_{sp} for Drained Load Tests

Data	Statistics	Q_{scm}/Q_{scp}	Data	Statistics	Q_{sum}/Q_{sup}
	n	37		n	27
DCR	mean	2.25	DUR	mean	1.64
	COV	0.37		COV	0.53
	n	44		n	31
DCS	mean	2.44	DUS	mean	1.88
	COV	0.31		COV	0.50
	n	81		n	58
All data	mean	2.35	All data	mean	1.77
	COV	0.34		COV	0.51

Comparison of compression and uplift side resistances is shown in Figure 6. For comparison, round and square piles were combined since their behavior is somewhat comparable. Results of all data combined for compression and uplift are likewise presented in Table 12. The mean side resistance ratios are 2.35 and 1.77 for compression and uplift load tests, respectively. The COVs are 0.34 and 0.51 for compression and uplift, respectively. The regression lines shown in Figure 6 indicate that $Q_{scm} = 1.96\ Q_{scp}$ and $Q_{sum} = 1.51\ Q_{sup}$ and are in good agreement with the mean results. Apparently, compression loading exhibits greater underestimation of side resistance than uplift loading.

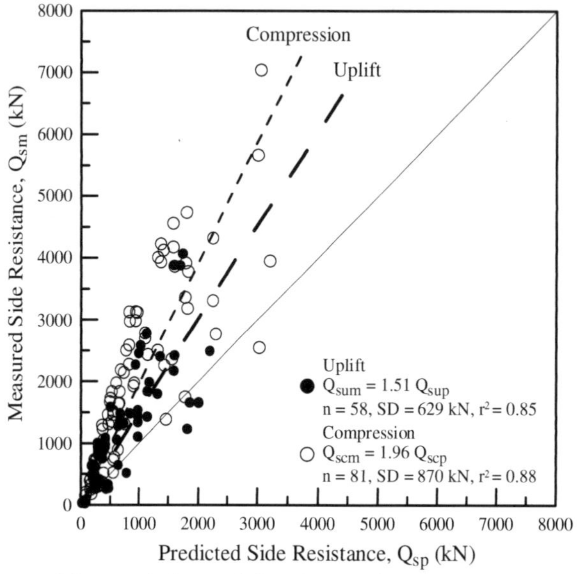

Figure 6. Q_{sp} and Q_{sm} for Drained Loading

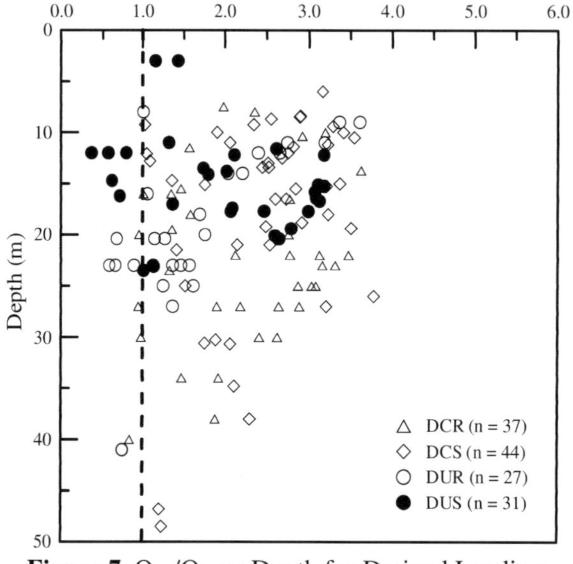

Figure 7. Q_{sm}/Q_{sp} vs Depth for Drained Loading

The underprediction of side resistance can be attributed to several factors. Underestimation of soil parameters is one possible reason because due to pile driving, the fact that the soil surrounding the pile becomes denser may have been neglected. Another reason may be due to the overconsolidation at shallower pile depths in which K_o may have been underestimated. The stress coefficient K/K_o which was assumed to be equal to 1.0 may be another reason.

The effects of pile depth on side resistance are examined in Figure 7. In general, the behavior for compression and uplift is comparable. It can be observed that the ratio of Q_{sm}/Q_{sp} generally decreases as the depth of the pile increases. However, for uplift square piles, the depth range is small and a wide variation is observed from the data points. The β method appears to be more consistent for long piles because it shows a wide range of results for short piles. In general, this phenomenon supports the above analysis that K_o has been underestimated at shallower depths.

To verify the issue of underestimation of side resistance in drained soils, K/K_o values were back-calculated utilizing the field load test data. As a first approximation, the measured beta (β_m) can be computed as follows:

$$\beta_m = Q_s(L_2)/[pD\bar{\sigma}_{vm}] \tag{6}$$

In which Q_s (L_2) = interpreted side resistance from L_2 method and $\bar{\sigma}_{vm}$ = mean vertical effective stress. Using Eqs. 5 and 6 and with the assumptions of $\beta_m = \beta_p$ and $\delta/\phi = 1.0$, the mean K/K_o was back-calculated for the overall foundation depth of driven PC piles. Table 13 lists the statistical results for K/K_o. For compression, the K/K_o values are 1.98 for round piles and 2.15 for square piles. For uplift, the K/K_o values are 1.47 for round piles and 1.67 for square piles. Square piles yield somewhat larger stress factor which can be attributed to the larger perimeter of a square pile for a same area of a round pile. The larger perimeter can provide larger influence area of denser soil resulting to better side resistance.

The K/K_o values for drilled shafts were recommended by a previous study [4]. It is suggested that for drained tests, K/K_o are 0.73, 0.97, and 1.03 for slurry, casing, and dry construction respectively, whereas for undrained tests, K/K_o are 0.79, 0.88, and 1.12 for slurry, casing, and dry construction respectively. Comparison of these values with the K/K_o values developed for driven piles indicates that larger stress factor can be expected for driven PC piles. This can be attributed to the installation method of driven PC piles to which the driving procedure provides denser soil surrounding the pile.

Table 13. Back-Calculated K/K_o

Test type	Pile section	Statistics	K/K_o
Compression	DCR	n	37
		mean	1.98
		COV	0.37
	DCS	n	44
		mean	2.15
		COV	0.31
Uplift	DUR	n	27
		mean	1.47
		COV	0.54
	DUS	n	31
		mean	1.67
		COV	0.50

3.2.2. Undrained load tests. A similar evaluation is done for undrained β analysis. The detailed analysis results for undrained load tests are presented in Tables 4 and 5 and Tables 8 and 9 for compression and uplift loading and are summarized in Table 14 including the statistics. The mean side resistance ratios (Q_{scm}/Q_{scp}) are 1.05 and 1.04, for round and square cross section piles, respectively. The COVs for round piles is 29% while it is 30% for square piles. The mean predicted side resistance is in quite good agreement with the mean measured side resistance. For uplift loading, the mean side resistance ratios (Q_{sum}/Q_{sup}) are 0.86 and 0.88, for round and square cross section piles, respectively. The COVs for both sections are 30%. The mean predicted side resistance is in somewhat good agreement with the mean measured side resistance. Contrary to drained load tests, it appears that the β method reasonably predicts the undrained side resistance of driven precast concrete piles.

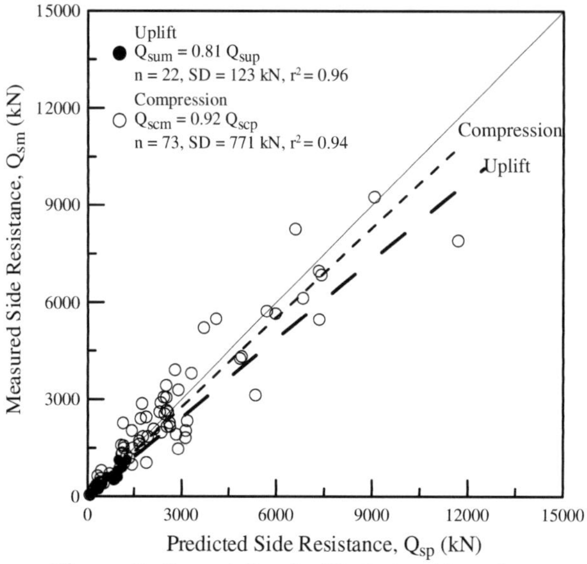

Figure 8. Q_{sp} and Q_{sm} for Undarined Loading

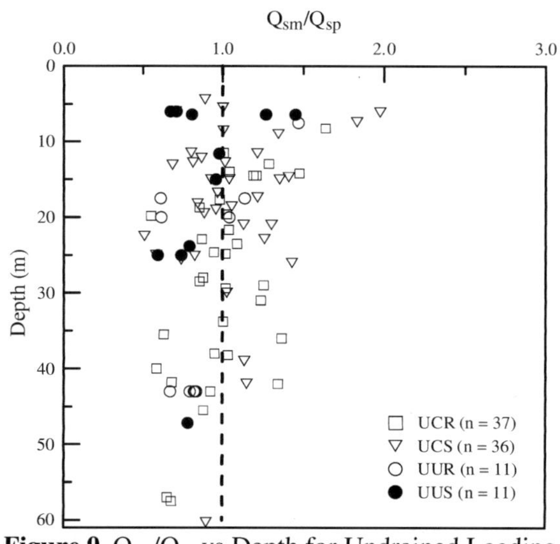

Figure 9. Q_{sm}/Q_{sp} vs Depth for Undrained Loading

Comparison of undrained compression and uplift side resistances is shown in Table 14 and Figure 8. Similar to drained tests, round and square piles were combined because their behavior is somewhat comparable. Results of all data combined for both compression and uplift in Table 14 indicate that the mean side resistance ratios are 1.05 and 0.87 for compression and uplift load tests, respectively. The COVs are 0.29 and 0.30 for compression and uplift, respectively. The regression analysis shown in Figure 8 indicates that Q_{scm} = 0.92 Q_{scp} and Q_{sum} = 0.81 Q_{sup}. Uplift loading exhibits a slight overprediction of side resistance. However, in general, β method can reliably be used in undrained side resistance analysis of driven PC piles.

As in drained loading, examination on the effects of pile depth to side resistance for undrained condition is explored and is shown in Figure 9. In general, the scatter is substantial for undrained loading. The effect of depth on side resistance is not explicitly defined by the data points due to a wide range of results throughout the depth with Q_{sm}/Q_{sp} ranging from 0.5 to 1.50. Although it can also be noted that wider range (0.50 to 2.0) is observed for shallower depths (< 10 m). Hence, careful engineering judgment on the use of the β method in undrained conditions is suggested.

Table 14. Statistics of Q_{sm}/Q_{sp} for Undrained Load Tests

Data	Statistics	Q_{scm}/Q_{scp}	Data	Statistics	Q_{sum}/Q_{sup}
	n	37		n	11
UCR	mean	1.05	UUR	mean	0.86
	COV	0.29		COV	0.30
	n	36		n	11
UCS	mean	1.04	UUS	mean	0.88
	COV	0.30		COV	0.30
	n	73		n	22
All data	mean	1.05	All data	mean	0.87
	COV	0.29		COV	0.30

3.3. λ method

Undrained load tests were evaluated using λ method. The value of λ for driven PC piles was back-calculated from field load test results as follows:

$$\lambda = \frac{Q_s(L_2)}{[pD(\bar{\sigma}_{vm} + 2s_u)]} \qquad (7)$$

In which all terms have been defined previously. Figures. 10(a) and 10(b) demonstrate the variation of λ to pile depth for compression and uplift loading, respectively. Both figures show a somewhat wider scatter for square piles than for round piles for shorter pile depths (D < 30m). The combined round and square result presents the mean λ values for specified depths in the inclusive table. Result shows that λ generally decreases with increasing depth. Since the behavior of compression and uplift data is comparable, these data were merged to provide a relation applicable to both test types. The relation is shown in Figure 10(c) indicating a range of λ value for shorter piles (< 30 m) of 0.29 to 0.20 whereas for longer piles (> 30 m), λ value is ranging from 0.20 to 0.11. The variation of λ values can be adopted for side resistance analysis of driven PC piles.The λ method was likewise adopted by previous research [4] for drilled shaft in cohesive soils. Their findings indicated that λ method produces less reliable results when applied to drilled shafts. Comparison of the previous study [4] and this study reveals that lambda method is more applicable to driven piles than drilled shafts.

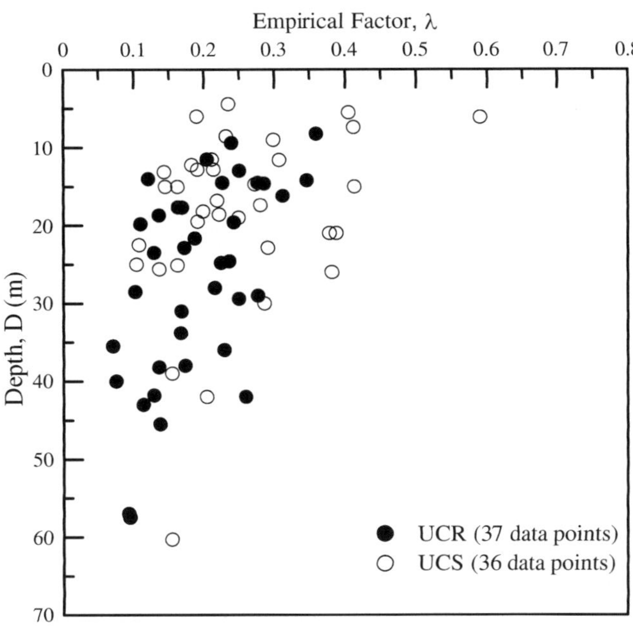

0-10m	n	9
	mean	0.33
	SD	0.13
	COV	0.38
11-20m	n	30
	mean	0.23
	SD	0.07
	COV	0.29
21-30m	n	18
	mean	0.22
	SD	0.10
	COV	0.43
31-40m	n	8
	mean	0.15
	SD	0.05
	COV	0.36
41-50m	n	5
	mean	0.14
	SD	0.04
	COV	0.28
51-60m	n	3
	mean	0.11
	SD	0.03
	COV	0.31

(a)

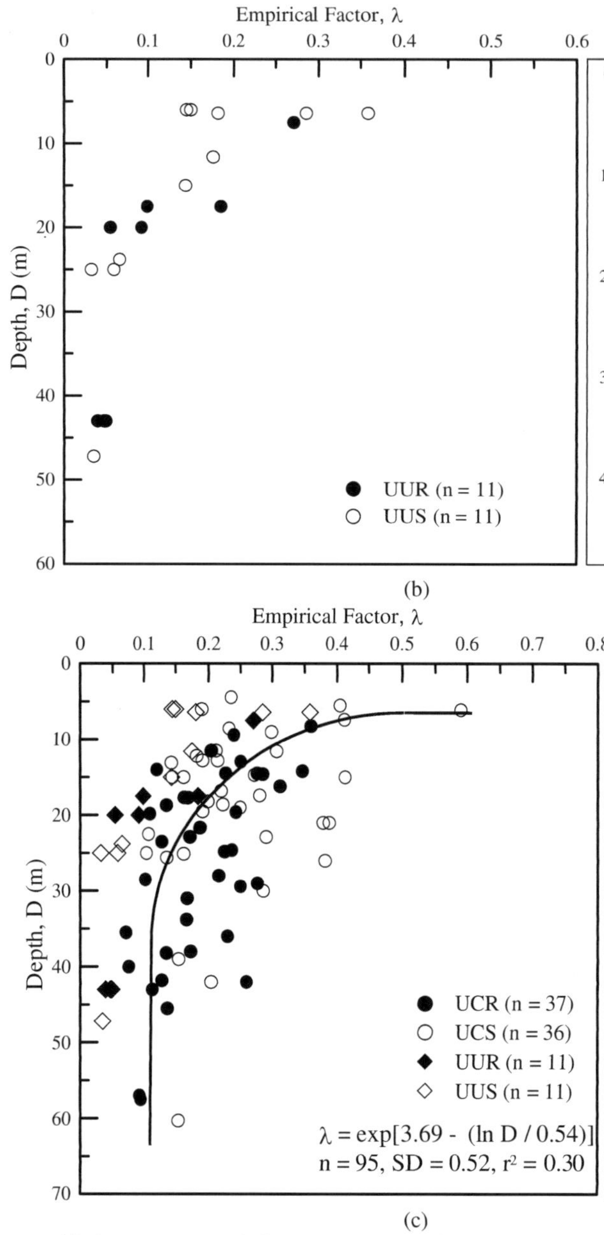

0-10m	n	6
	mean	0.23
	SD	0.09
	COV	0.37
11-20m	n	6
	mean	0.12
	SD	0.05
	COV	0.41
21-30m	n	3
	mean	0.05
	SD	0.02
	COV	0.34
31-40m	n	0
	mean	-
	SD	-
	COV	-
41-50m	n	7
	mean	0.04
	SD	0.01
	COV	0.13

0-10m	n	15
	mean	0.29
	SD	0.12
	COV	0.41
11-20m	n	36
	mean	0.21
	SD	0.08
	COV	0.37
21-30m	n	21
	mean	0.20
	SD	0.11
	COV	0.54
31-40m	n	8
	mean	0.15
	SD	0.05
	COV	0.36
41-50m	n	12
	mean	0.11
	SD	0.07
	COV	0.68
51-60m	n	3
	mean	0.11
	SD	0.03
	COV	0.31

Figure 10. λ versus Depth for (a) Compression (b) Uplift Loading, and (c) All Data

4. Conclusion

Compression and uplift field load test data were utilized to evaluate the side resistance of driven precast concrete piles. For drained loading, β method was applied while for undrained loading, α, β, and λ methods were applied. Based on these analyses, the following conclusions emerge.

- For undrained loading, the α_{CIUC} - $s_u(CIUC)/p_a$ correlation is developed using field load test data and can be utilized for driven PC pile total stress analysis. For undrained loading, the correlation α_{CIUC} - $s_u(CIUC)/\bar{\sigma}_v$ - $\bar{\sigma}_v$ developed from field load test data can be regarded as an alternative base for driven PC pile total stress analysis.
- For drained loading, β method undepredicts the side resistance. The method is more consistent for long piles because it shows a wide range of results for short piles. For undrained loading, β

method reasonably predicts the side resistance. The suggested stress factor K/K_o for drained compression loading is 1.98 for round piles and 2.15 for square piles whereas for drained uplift loading, the suggested stress factor K/K_o is 1.47 for round piles and 1.67 for square piles. These values can substantially improve the pile capacity prediction using β method.

For undrained loading, the compression and uplift λ versus depth relations for driven PC piles are developed that can be utilized for pile analysis and design. The α, β, λ methods can reasonably be applied in driven piles under undrained loading conditions, whereas, β method can be suitable for driven piles in drained loading condition with the use of appropriate stress factors.

Acknowledgments
This study was supported by the Center for Research and Development, Adamson University and the National Science Council, Taiwan, under contract number: NSC 100-2221-E-033-073-MY3.

References

[1] O'Neill M W 2001 *J. Geotech. and Geoenvi.* Eng. **127** pp 1 – 16
[2] Lutenegger A J 2001 *International foundation congress and equipment expo,* ASCE Florida pp 9 – 17
[3] Chen Y J and Kulhawy F H 1994 *Case history evaluation of behavior of drilled shafts under axial & lateral loading Final Report TR-104601* (Palo Alto, California: Electric Power Research Institute)
[4] Chen Y J, Lin S S, Chang H W and Marcos M C 2011 *J. Mar. Sci. and Tech.*, **19** pp 210 – 221
[5] Tomlinson M J 1957 *4th International conference on soil mechanics and foundation engineering* **2** pp 66 – 71
[6] Stas C V and Kulhawy F H 1984 *Critical evaluation of design methods for foundations under axial uplift and compression loading Report EL-3771* (Palo Alto, California: Electric Power Research Institute)
[7] Randolph M F and Murphy B S 1985 *17th Offshore technical conference* Houston **1** pp 371 – 378
[8] Semple R M and Rigden W J 1986 *Grnd. Eng.* **19** pp 11 – 17
[9] Kulhawy F H and Jackson C S 1989 *Proc. Foundation engineering: Current principles and practices* USA pp 1011 – 1025
[10] Kulhawy F H, *Foundation engineering handbook (2nd Ed)* 1991 (New York: Van Nostrand Reinhold) pp 537 – 552
[11] Kulhawy F H, Trautmann C H, Beech J F, O'Rourke T D, McGuire W, Wood W A and Capano C 1983 *Transmission line structure foundations for uplift - compression loading Report EL-2870* (Palo Alto, California: Electric Power Research Institute)
[12] Vijayvergiya V N and Focht J A jr 1972 *4th Offshore technology conference* pp 865 – 874
[13] Kulhawy F H and Mayne P W 1990 *Manual on Estimating Soil Properties for Foundation Design Report EL-6800* (Palo Alto, California: Electric Power Research Institute)
[14] Hirany A and Kulhawy F H 1988 *Conduct & interpretation of load tests on drilled shaft foundations: Detailed guidelines Report EL-5915(1)* (Palo Alto, California: Electric Power Research Institute)
[15] Hirany A and Kulhawy F H 2002 *Deep foundations (GSP 116)* ed M W O'Neill & F C Townsend (Reston: American Society of Civil Engineers) pp 1018 – 1028
[16] Chen Y J and Fang Y C 2009 *J Geotech. and Geoenvi. Eng.* **135** pp 1056 – 1069
[17] Marcos M C, Chen Y J and Kulhawy F H 2013 *KSCE J. Civil Eng.* **17** pp 1 – 15
[18] Marcos M C and Chen Y J 2018 *IJETI* **8** pp 118 – 132
[19] Marcos M C 2013 *Ph.D. Dissertation* Chung Yuan Christian University, Taiwan pp 101-102
[20] Chen Y J and Kulhawy F H 1993 *J. Geotech. Eng.* **119** pp 1732 – 1750

IOP Conf. Series: Materials Science and Engineering **658** (2019) 012006 doi:10.1088/1757-899X/658/1/012006

Solar radiation prediction based on random forest of feature-extraction

J Liu[1], M Y Cao[2], D Bai[1] and R Zhang[1]

[1]College of Information and Control Engineering,Shenyang Jianzhu University, Shenyang 110168, China
[2]Qianju Technology Development Co., LTD, Beijing 100094, China

E-mail: jeanliu10@163.com; 1582729192 @qq.com

Abstract: In order to use solar energy efficiently, the accurate prediction of solar radiation is extremely important. The aim of this paper is to discuss and build the solar radiation prediction model, and an efficient method based on improving random forest is proposed in here. The random forest regression model is created for predicting solar radiation. The nine variables are taken as input of the original model, and the solar radiation as the output. The original data is normalized using Min-Max method and is mapped to the values in the range of zero to one. The large amount of radiation data has been dealt with univariate feature extraction. The classifier of the random forest tree is constructed by characteristic factors that can influence solar radiation, and the optimal parameters of the model are selected by the OOB error analysis. The relationship between M-try and OOB can be obtained by computing of the R platform. Hence, a prediction model of improved random forest solar radiation based on feature extraction has been described. Under the specific area of solar radiation forecast, the studies have been done, and the solar radiation is analyzed and compared with the different prediction methods. The simulation results prove the validity of the improved method; the root mean square error is reduced and the precision of prediction is improved. It is significant for the prediction of the amount of solar radiation under complex environment and the effective use of photovoltaic power generation.

1. Introduction

With the increase of environmental pollution, in recent years, solar energy as optimal clean new energy resources have attracted more and more attention. So, how to forecast accurately the solar radiation is particularly critical for the solar energy utilization [1-8].

In the first instance, Angstrom proposed the climatology method which calculates monthly total solar radiation through utilization of solar radiation under clear day and the sunshine percentage [9]. Besides, Zhou Jin and some others fitted the coefficient of the correction formula according to the solar radiation data from 69 meteorological stations and the sunshine material [10]. In recent years, more and more intelligent prediction methods have been applied in the solar radiation prediction. For example, Zhang Jingjing [11] established a BP neural network simulation model to predict the solar radiation, which was trained by using the actual data and realized the prediction of future solar radiation. Many researchers have proposed solar radiation prediction based on SVM and SVR. Such as Zen Jianwu and Qiao Weiwho built a Least Squares Support Vector Machine (LS-SVM) model to predict the amount of solar radiation in the short term [12]. As for the strong non-linearity and stationary of the daily solar radiation energy time series, Wang Jianping applied neural network combining with the Fourier transform and wavelet multi-scale decomposition to equivalently establish daily solar radiation energy wavelet process neural network real-time online prediction model [13], which has more degrees of freedom and a stronger nonlinear approximation and random estimation ability, especially influenced by many factors such as atmospheric condition, solar elevation, Earth-Sun distance, the sunshine time, the altitude, terrain and landforms, the obstacles and so on. This paper mainly studies the solar radiation prediction problem which is based on the random forest algorithm, because the accurate and effective solar radiation prediction method has vital actual significance to develop solar energy resources and improve the efficiency of photovoltaic power generation effectively [14, 15].

The contents of the paper consist of five sections. In section 2, the random forest algorithm is introduced. The solar radiation prediction model is established based on random forest algorithm. In section 3, we analyze the parameters with the amount of solar radiation and then propose the solar radiation predication model based on feature extraction of random forest. In section 4, the improved method based on the characteristics processing has been compared with the original one, in case of the monthly average of solar radiation value. In section 5, the research conclusions are presented.

2. Random forest algorithm

Random forest (referred to as RF) [16-18] is composed of many decision tree model $h(X;\theta_k)$, $k=1,2,...,K$ which is a combined regression model. Where θ_k are independent and identically distributed random vectors. Let X be the input and Y be the output variables, then (X, Y) constitute the original data set. The construction of each decision tree includes two processes of sampling and complete split. The first step constitutes of two random sampling processes, in which the input data of random forests are separately made for row sampling and column sampling. In row sampling, taking back method is used, and namely repeated samples may be contained in the sampled data sets. Assume that the input sample is N, then sampling samples is n (n<N). So, in training, the input samples of each tree are not all the samples. On the column sample, we randomly select m (m << M) from the M attributes. Then using the split of the data sampling method, we establish decision tree.

Since the two random samples ensure the randomness of the aforementioned process, then there is no need for pruning, and it will not appear over-fitting [19]. Random forest method does not generate over-fitting as the decision tree increases. The generalization error tends to be upper bound. We give the theoretical basis as the following.

Definition 1: Let $I(.)$ and $aV_k(.)$ are two functions for expressing average value. $mg(x,y)$ is the margin function of the average values. Y is correctly classified for the vector X. The last term represents maximum value of votes that vector X is classified by mistake. The marginal function reflects the degree of difference between the two. The higher the value, the combination classifier performs better. It is represented as follows:

$$mg(x,y)=aV_k I(h(x;\theta_k)=y)-\max_{j\neq y} aV_k I(h(x;\theta_k)=j) \tag{1}$$

Definition 2: Assume that generalization error reflects the classification ability of samples, which does not appear in the training set. The smaller the value is, the better the performance of the algorithm gets. The generalization error PE^* is defined as follows:

$$PE^* = P_{x,y}(mg(x,y)<0) \tag{2}$$

Theorem 1: When the random number is large in forest tree, the tree structure and the law of large numbers to ensure that the random forest does not produce over-fitting:

$$\lim_{k\to\infty} PE^* = P_{x,y}(P_\theta h(x,\theta_k)=y)-\max_{j\neq y}(P_\theta h(x,\theta_k)=j) \tag{3}$$

Theorem 2: The generalization error that exists on the boundary is given by the following equation:

$$PE^* = \rho(1-s^2)/s^2 \tag{4}$$

Where, ρ reflects the correlation between decision trees. S reflects the ability to classify single decision tree. Thus, from equation (4), we can know that the smaller the correlation between decision tree, the better the single trees classification ability of decision tree, and the less generalization error in random forests.

While, Gini Index is adopted in the splitting attribute of decision tree, assuming that T contains N categories of records, hence the index is:

$$Gini(T) = 1 - \sum_{j=1}^{n} P_j^{\,2} \tag{5}$$

Whereis category, j is the frequency of occurrence.

If set the T is divided into the m partsN1, N2 ,… Nm, then the partition Gini is:

$$Gini_{split}(T) = \frac{N_1}{N} Gini(T_1) + \ldots + \frac{N_m}{N} Gini(T_m) \tag{6}$$

By the k rounds training, we get a regression Prediction model list $\{h(X;\theta_1), h(X;\theta_2), \ldots, h(X;\theta_k)\}$. Then a combined forecast model is built by these lists. The final result is:

$$\bar{h}(X) = (1/K)\sum_{k=1}^{K} h(X;\theta_k) \tag{7}$$

The training set of the Random forest has been generated by means of bagging method. The Out-Of-Bag (referred as OOB) generated with each sampling data is used to evaluate the accuracy of random forest regression prediction.

3. Prediction model based on feature extraction of random forest

3.1. Parameters with the amount of solar radiation

Solar radiation is the main energy of the earth's atmospheric motion. The total radiation intensity includes direct and the scattered radiation intensity. In the case of atmospheric attenuation, if the direct solar radiation at some point is E_b and the direct solar radiation on a horizontal plane is E_0, then the formula of E_b and the direct radiation intensity E_{bh} which consider solar elevation angle were:

$$E_b = E_0 \times \tau_b \tag{8}$$

$$E_{bh} = E_0 \times \tau_b \times \sinh \tag{9}$$

There, τ_b is a direct radiation atmospheric transparency coefficient; h is the height of the sun angle.

As sun scattering part is more complex, it has some connection with cloud amount, atmospheric, cloud-shaped distribution of the sky. Let E_d be the diffused solar irradiance to arrive on horizontal plane, then:

$$E_{dh} = \frac{1}{2} \times \sinh \times \frac{1-\tau_d}{1-1.4\ln[\tau_d / m(z,b)]} \times k \tag{10}$$

Where, τ_b is the scattered radiation atmospheric transparency, k is selected according to the quality of air in between 0.60 and 0.90.

In general, the intensity of solar radiation reaching the ground can be E_{th}:

$$E_{th} = E_{bh} + E_{dh} \tag{11}$$

3.2. Random forest forecast model based on feature-extraction

Feature extraction is a process for the related amount of solar radiation data component decomposition, recombination and selection. It is a key part of random forests, which not only determines the quality of subsequent processing, efficiency, system complexity and robustness, but also to be able to dig what knowledge and the results of the physical meaning of interpretability. We use the single variable feature extraction algorithms to calculate the relationship between a feature and regression variables, so that to calculate a score for each feature, excluding those important indicators. Here, a chi-square method to test features score is applied.

Chi-square test is a common distribution hypothesis testing method based on χ^2. The null hypothesisH0is that there is no difference observed in frequency and expected frequency. The basic idea of the test is that we assume H0is correct, the valueχ^2 is calculated based on this premise. That indicates that the degree of deviation from the observed values with the theoretical value. According to the Chi-square distribution and degrees of freedom we can correctly determine the current statistics in the case of H0 and the probability under more extreme circumstances. If the P value is small, the degree observation value with the theoretical value is too big, hence we should reject the null hypothesis.

Considering the difference between the order of magnitude of these variables is larger, in order to make the model predictions more accurate, we have normalized the original variables, which have been transformed into dimensionless data. Using the Min-Max normalization method, all the data is mapped to the original values in the range of zero to one. The formula of the normalized is as follows:

$$x_i^* = \frac{x_i - x_{\min}}{x_{\max} - x_{\min}} \tag{12}$$

Where, x_i^* is the input variable after been normalized, $x_{\min}$ and $x_{\max}$ are minimum and maximum values of the original data, respectively. The normalized vectors are used as input of the RF algorithm, corresponding to the amount of solar radiation for the output of the RF algorithm.

Here, the data is the amount of solar radiation of a four-year 48 sets. The data formats Random Forests are analyzed in the database n×m. Each row is used as a sample of the data and m-1 columns is used as variables. Then we create an improved random forest regression prediction model of solar radiation with nine inputs and one output. The random forest regression models are structured thought MATLAB.

4. Simulation results and analysis

4.1. Prediction model simulation and comparison

To verify the feasibility of study, using a common solar radiation monitoring database data from laboratory of Shenyang Institute of Engineering is used. Taking the Shenyang ($41^0 50' N$,$135^0 28' E$) 2004 - 2014 solar radiation data for a sample, which includes sunshine hours, average atmospheric pressure, average wind speed, the average air temperature, solar elevation angle, the average humidity.Table1 lists the predicted result of the three kinds of model of 2014. Due to the randomness of daily solar radiation at different times, discrete data can not completely describe the solar radiation principle, monthly average solar radiation value can better reflect the trend of solar radiation and the affected factors. Therefore, the monthly average of solar radiation value has been taken as in our research. Then we regard all data as monthly average value in data processing.

Table 1. Predicted result of the three kinds of model of 2014($MJ \cdot m^{-2} / month$).

Month	MV	IRF	RF	SVM
1	114.32	96.99	88.18	83.65
2	135.74	160.52	108.39	157.47

3	201.38	223.04	218.16	172.00
4	316.04	297.83	288.00	287.93
5	298.55	321.64	315.90	320.91
6	374.63	368.28	349.99	343.22
7	403.58	383.87	422.65	371.55
8	410.72	430.77	387.40	428.07
9	365.91	338.05	340.34	332.88
10	355.38	339.34	324.71	326.05
11	217.67	203.31	231.75	237.83
12	128.51	111.33	100.35	97.10

In order to further verify the validity of this model, the real data was predicted through the experiment. Here, we predicted the monthly total solar radiation in April, July, October, January ranging from the 2004 to 2013 with the same three methods, with four month forecast data as four seasons. The solar month radiation predicted results with three kinds of methods can be seen in (a), (b), (c), (d) of the figure 1.

In addition, under the same training samples from 2004 to 2013, the three kinds of solar radiation average radiation prediction model are established, respectively. The predicted results of the solar year radiation are as shown in figure 2. From the fig.2, the prediction effect of the improved random forest method is better than others.

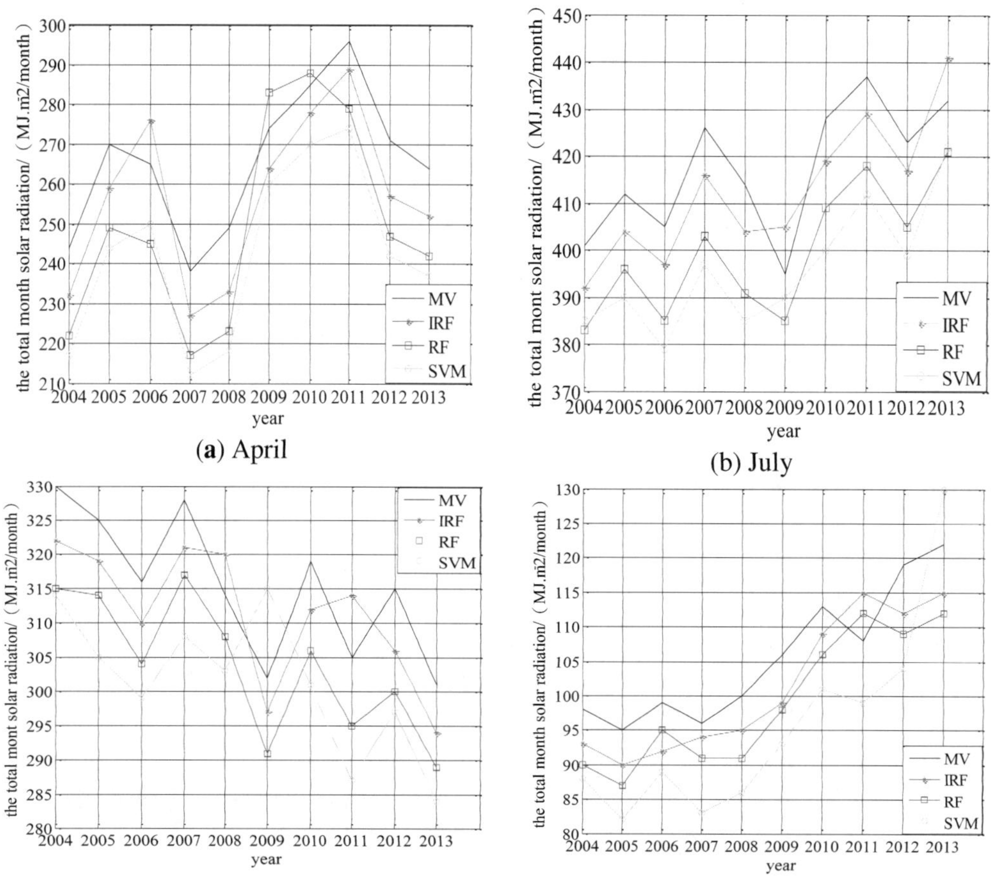

(**a**) April (b) July

(c) October (d) January

Figure 1. Solar month radiation predicted results in different months.

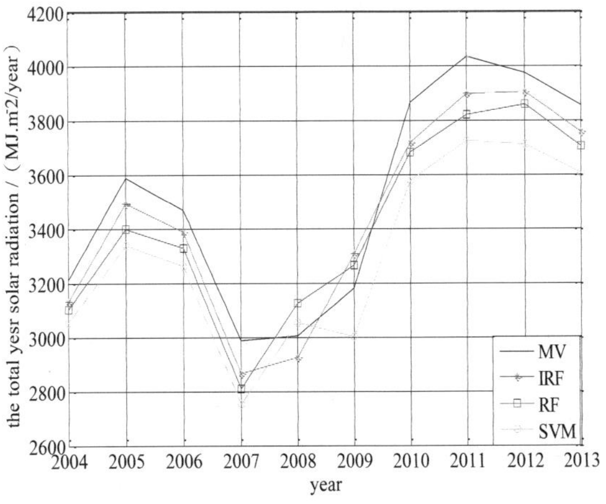

Figure 2. Solar year radiation predicted results.

4.2. Model evaluation and inspection

It is a core that the evaluation and test of model in the entire modeling process, through model test to determine the fitting effect of the model. The model is evaluated based on the model decision coefficient (R2) and root mean square error (RMSE).

$$R^2 = 1 - \frac{\sum_{i=1}^{n}(y_i - \hat{y}_i)^2}{\sum_{i=1}^{n}(y_i - \bar{y})^2} \tag{13}$$

$$RMSE = \sqrt{\frac{\sum_{i=1}^{n}(y_i - \hat{y}_i)^2}{n}} \tag{14}$$

Where, y_i is the measured value; $\hat{y}_i$ is the model forecast; $\bar{y}$ is the average value of the sample; n is the sample.

Through testing of the improved random forest regression model of the samples, the deviation statistics and evaluation of the predictive ability of model, have been compared through some statistics which include the general relative error (Rs), the average relative error (E1), the overall relative error absolute value (E2) and precision (P) are as follows:

$$R_S = \frac{\sum y_i - \sum \hat{y}_i}{\sum \hat{y}_i} \tag{15}$$

$$E_1 = \frac{1}{n}\sum(\frac{y_i - \hat{y}_i}{\hat{y}_i}) \times 100\% \tag{16}$$

$$E_2 = \frac{1}{n}\sum\left|\frac{y_i - \hat{y}_i}{\hat{y}_i}\right| \times 100\% \tag{17}$$

In the formula, y_i is the observed values; $\hat{y}_i$ is the model forecast; n is the predicted sample.

The sample mean and sample can be calculated based on the sun's total radiation measured values in 2014, taking the results both of the calculation and the prediction into the formula (13) and (14). The accuracy test result as shown in table 2. Now, the decision coefficient of random forest regression model $R^2 = 0.9665$, the root means square error $RSME = 19.6085$, the fitting effect of this model is ideal and better than the other.

Table 2. Accuracy test results of the three kinds of model.

Month	R2	RMSE
IRF	0.9665	19.6085
RF	0.9448	23.9800
SVM	0.9329	27.724

The improved random forest regression model has the highest forecast precision, it can be seen that the improved random forest model forecast ability is better.

5. Conclusions

In this paper, an improved random forest algorithm is presented. The decision trees have been constructed and the variables of importance ranking have been optimized by evaluation function of random forest function. It is assumed that under the same condition, the three kinds of method have been compared. Studies and trials have led to the conclusion that under supposed conditions of too many cloudy days, the simulation conditions, using the improved random forest model predicted better than the results of other two algorithms. The new method is significant for the prediction of the amount of solar radiation under complex environment and for the use of photovoltaic power generation. There are a number of factors not taken into consideration, such as topography, altitude factor and so on, this segment still requires to be further researched.

Acknowledgments

This work is supported by the scientific research projects in Liaoning Province Natural Science Foundation of China （201602616），Liaoning Province Department of Education (L2015443) and Ministry of Housing and Urban-Rural Development of China (2015-K2-015).

Author contributions

For the research articles, the four authors distributed the responsibilities as Liu Jian carried out the theoretical algorithm research; Cao Meiyan carried out the experimental research and perfumed the analysis; Bai Di performed the experiments; Zhang Rui conducted the experimental data acquisition and analysis; Liu Jian and Cao Meiyan wrote the paper.

References

[1] Yadav AK and Chandel SS 2014 *J. Renewable and Sustainable Energy Reviews* **33** pp 772－781
[2] Yang MQ and Wang XY 2014 *J. Applied Mechanics and Materials* **571** pp 217－222
[3] Dıaz F, Montero G and Escobar J M 2014 *J. Main Thematic Areas* **69**
[4] Alonso-Montesinos J and Batlles F J 2015 *J. Energy* **83** pp 387－393
[5] Feudo T L,Avolio E and Gullì D 2015 A preliminary study, in a coastal site of South Italy (Lamezia Terme), **17** p 11774
[6] Voyant C, Darras C and Muselli M 2014 *J. Applied Energy* **114** pp 218－226
[7] Troccoli A and Morcrette JJ 2014 *J. Journal of Applied Meteorology and Climatology* **53** pp 2571－2588
[8] Peng H W, Yang X F and Liu F R 2010 *J. Power System and Clean Energy* pp 82－84
[9] Angstrom A 1924 *J. Journal of the Royal Meteorological Society* **50** pp 121－126
[10] Zhou J, Wu Y Z and Yan G 2006 *J. Journal of Harbin Institute of Technology* **38** pp 925－927
[11] Zhang JJ 2014 *J. Industrial control computer* pp 39－40
[12] Zeng J and Qiao W 2013 *J. Renewable Energy* **52** pp 118－127
[13] Wang J, Jiang H and Wu Y 2015 *J. Energy* **81** pp 627－644
[14] Hosenuzzaman M, Rahim N A and Selvaraj J 2015 *J. Renewable and Sustainable Energy Reviews* **41** pp 284－297
[15] Kaushika N D, Tomar R K and Kaushik S C 2014 *J. Solar Energy* pp 327-342

[16] Gu W, Vieira A R and Hoekstra R M 2015 *J. Epidemiology and Infection* pp 1-9
[17] Ramedani Z, Omid M and Keyhani A 2014 *J. Renewable and Sustainable Energy Reviews* **39** pp 1005 – 1011
[18] Liaw A and Wiener M 2002 *J. R news* **2** pp 18 – 22
[19] Criminisi A, Shotton J and Konukoglu E 2012 *Foundations and Trends in Computer Graphics and Vision* **7** pp 81 – 227

Tree of life – intelligent environment system station design

Po-Chuan Chen[1] and Jiunn Fang[2]

[1]Program of Mechanical and Aeronautical Engineering, Feng Chia University, Taichung City, Taiwan
[2]Department of Aerospace and Engineering, Feng Chia University, Taichung City, Taiwan

Email: chenname0317@gmail.com

Abstract. The three elements essential for life are sunshine, air, and water. This is an undeniable fact, but with advancements in technology, amenities such as electricity, smart phones, and the internet have become the urban versions of these essential elements for life. Most people living in cities desire constant connectivity. This connectivity requires significant electricity and access to a high-speed internet connection. The purpose of this research is to design a wind turbine system modelled after the shape of a tree with the blades built inside the tree. The power generated by the tree will be stored in a power bank located inside a station adjacent to the tree and will be available to the public. Additionally, there will be WLAN antennas built inside the tree branches to provide wireless internet signals. Finally, the electricity produced will be green energy, which will help us achieve the goal of eco-friendly but intelligent life.

1. Introduction

Our life has changed drastically since the smart phone appeared. In the past, the cell phone only has simple functions like sending and receiving calls or text messages. Today, smartphones can replace many things like wallets, tickets or even car keys. Therefore, power has become a big issue for phone producing companies. Still, no matter how big the battery is inside the phone, the power is never enough. Thus, a power bank seems to be everyone's must bring item. However, a power bank can be heavy and take up lots of space.

Moreover almost every function in the smart phone requires internet connectivity. For those who do not have unlimited access to a cellular internet, government provided free Wi-Fi can be very useful. The Tree of Life - Intelligent Environment System Station provides these two major functions: a green energy generated power bank and free public Wi-Fi.

1.1. Traditional wind turbines

A wind turbine is a device that converts wind energy into electrical energy. This device is composed of wide blades and is usually oriented along either vertical or horizontal axis. Different sizes of wind turbines provide different levels of power for various devices. The wind speed can be categorized in different levels of wind powers as shown in table 1.

Poul La Cour, a meteorologist from Denmark, invented the first wind turbine in the late 19th century. However, the low efficiency of these early wind turbines made them unpopular until recent years due to the rise in awareness of eco-friendly power.

During the 1980s, wind turbines were able to reach 55 kW of generated power. Today, due to advances in technology, wind turbines are able to reach a maximum power of 5000 kW, and there are more than 600,000 devices installed all over the world.

1.1.1. Horizontal axis wind turbine The rotating speed of horizontal axis wind turbines is high. Almost every horizontal axis wind turbine is equipped with a wind-facing device that allows the turbine to change direction to receive the wind. For large horizontal axis wind turbines, this kind of device uses sensors to detect the wind direction. The horizontal axis wind turbine requires a higher wind speed and energy to start, however, the voltage generates is also higher [1].

1.1.2. Vertical axis wind turbine Conversely, vertical axis wind turbines require lower wind speed to start the blade rotation, but because of the starting power requirements, the efficiency of vertical axis wind turbines is lower than horizontal axis turbines.

The wind restriction of this type of wind turbine is lower, meaning there will not be a need to face a certain direction to start the turbine. Further, owing to the vertical placement of vertical axis wind turbines, the space needed is much smaller than that of horizontal axis wind turbines. Therefore, it is more suitable for urban environments [2, 3].

Table 1. Performances of Wind Level

Level of the Winds	Corresponding Speed (m/s)		
0	0 ~ 0.2	7	13.9 ~ 17.1
1	0.3 ~ 1.5	8	17.2 ~ 20.7
2	1.6 ~ 3.5	9	20.8 ~ 24.4
3	3.4 ~ 5.4	10	24.5 ~ 28.4
4	5.5 ~ 7.9	11	28.5 ~ 32.6
5	8 ~ 10.7	12	>32.6
6	10.8 ~ 13.8		

1.2. Social benefits of wind turbine

The reason why renewable energy is more popular than conventional energy is because of its lower social costs. Traditional energy generation methods, such as coal-fired power generation, require nearly 40% more social cost per kwh of energy consumed than renewable energy.

The regenerative energy such as wind power generation not only reduces the persecution of the earth in the acquisition of raw materials, but also exerts lower pressure on the environment than traditional energy sources. Although the environmental pressure is low, it is not zero. The biggest shortcomings of wind turbines are two. The first is the noise generated by it and the second is the ecological impact. At present, the development of new-type wind power generation equipment has been very advanced, but the noise generated is a case of a one-megawatt wind turbine. The noise detected in a square of 300 meters is about 45 decibels.

At the same time, offshore wind power equipment is generally built on the windward coastline in order to obtain more wind energy. Usually, this area is also a route for general migratory birds, and many birds are also injured. However, compared with traditional energy generation, the environmental and social benefits provided by wind power generation are much higher, such as reducing environmental pollution, safe energy supply, lower cost in the long-term perspective, and even providing more employment opportunities. The creation of social prosperity, etc., these advantages have also made this power generation method popular in the world [4].

1.3. Public Wi-Fi service

By the end of 2018, the percentage of offload internet connections will reach another high mark due to the 16% year after year increase in smartphone and tablet usage. The reasons why people are choosing

IOP Conf. Series: Materials Science and Engineering **658** (2019) 012007 doi:10.1088/1757-899X/658/1/012007

to use Wi-Fi over cellular networks is the lower or even free cost of Wi-Fi. Consequently, the risks of using public Wi-Fi also rises. Therefore, people tend to use their mobile service more than offlaod traffic. This situation can be seen in Figure 1.

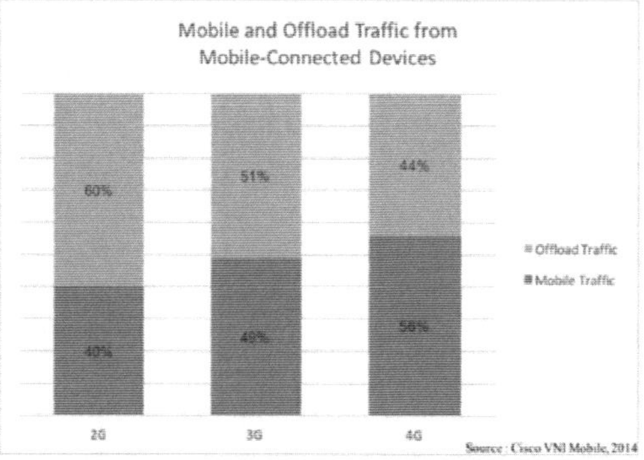

Figure 1. Percentage of Mobile and Offload Traffic from Mobile-Connected Devices

At the beginning of the twenty-first century, every major city in the world declared they would build public Wi-Fi for citizen and visitor use. Initially, this effort would seem to be very easy due to the mature technology of Wi-Fi service. However, it turns out not to be as easy as it seems because of security and blocking problems.

The major security problems of Wi-Fi are hacks. Because the majority of free Wi-Fi or public Wi-Fi network services are always unencrypted, anyone with cheap and readily available software can hack into any device connected to these unencrypted services. Therefore, wireless eavesdropping can happen on any free Wi-Fi service.

Even though people are aware of the safety issues of using public Wi-Fi, sometimes these issues are forgotten or ignored because they are in a rush or it is convenient. Therefore, connecting to a safe and certificated public Wi-Fi system is a better solution [5, 6].

2. Design and Methods

To achieve the goal of making an intelligent environment system station, this design took the three elements of life (sunshine, air, and water) as concepts and drew parallels with modern technologies. After analyzing the way people live, the results show that it is nearly impossible for people to live without any assistance from technology, especially as the Internet of Things has become so popular. Therefore, the three new elements of urban life are internet, electricity, and smart phone. This design combines these three entities into an intelligent environment system station with the hope of making people's lives easier.

2.1. Electricity system

To achieve better efficiency and save space, we will use H-type as the design concept. There will be three blades using NACA 0018 airfoil. The wind turbine will be two meters high and one meter in diameter. When the wind turbine is rotating in a fixed condition, the rated power is 783.2 W, as shown in Table 2.

Table 2. Wind Turbine Efficiency Analysis Parameter

Parameter	Value
Swept Area (M)	2
Rate Wind Speed (m/s)	12
Air Density (kg/m^3)	1.225
Power Factor Cp*	0.37
Rate Power (W)	783.2

2.1.1. Wind turbine blades design According to the result, NACA0018 airfoil has the best performance under low wind speed condition. Therefore, this design used this type of airfoil as blades [7]. The coordinate plot of NACA0018 can be seen in Figure 2.

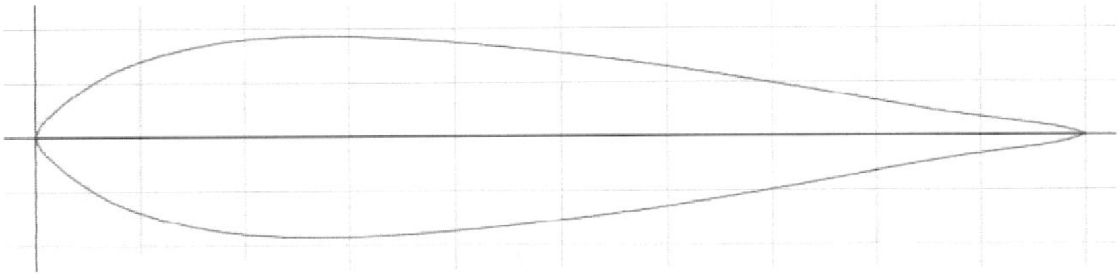

Figure 2. NACA0018 Airfoil

2.1.2. H-Type wind turbine structure To reduce the unnecessary spaces, this design use H-Type of wind turbine structure. This structure allows the design to achieve the goal of saving spaces but also generating high efficiency of electricity power as shown in Figure 3.

Figure 3. H-Type Wind Turbine Structure

2.1.3. Tree trunk structure design This research designed a tree trunk outside of the wind turbine as shown in Figure 4 to not only protect people from getting hurts by the rotating blades but also make this design more like a tree.

This tree trunk is three meters tall and the diameter is two meter, this size can entirely storage the whole wind turbine and keep one meter distance between the blades and the edge of the trunk. The main function of this tree trunk is to prevent people from injured, the ventilation calculation is not being tested in this period of research.

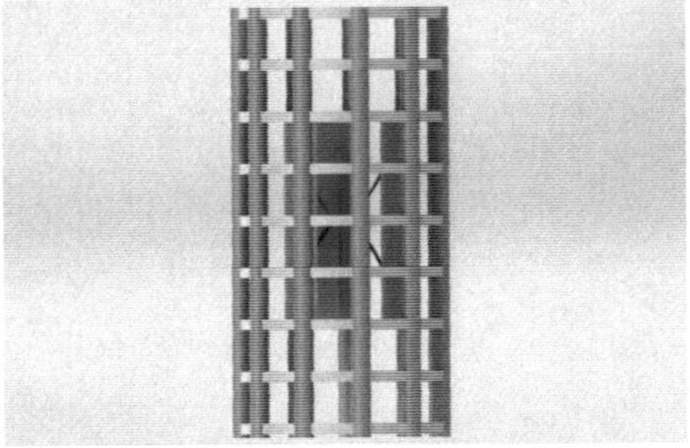

Figure 4. Tree Trunk Structure

2.2. Power bank land and return

The power generated by the wind turbine will charge the power banks located inside the station. The station design can be found in Figure 5. A touch screen installed inside the station allows people to register their personal information and start to land the power bank.

IOP Conf. Series: Materials Science and Engineering **658** (2019) 012007 doi:10.1088/1757-899X/658/1/012007

On the other side of the station, there is the area for battery to storage and charge and the image can be seen in Figure 6.

Figure 5. Smart Tree Station Design

Figure 6. Power Bank Storage Area

Every power bank has 5000 mAh power capacity as shown in Figure 7, which is almost enough to recharge a dead phone battery to 100% within an hour. On the back side of the battery, there is a special charging system installed. This unique charging system only allows the battery to only charge in this station, so when the battery is running out of the electricity, it can only be taken back to the station to charge. This unique charging system can be seen in Figure 8. Even though there are many other ways to bring a dead phone back to life. For example, there are many convenient stores where provides free charging services, but these services require users to stay at a certain location until their phones are fully charged. Also, bring a dead phone back to life only need few seconds, but if the purpose is to fully charge a dead phone require at least thirty minutes and above. Therefore, the portable power bank provides these benefits of convenience and mobility. When landing the power bank, the user swipes their metro card to pay the charge fee according to the amount of time the user landed. However, devices cannot be charged outside of the station so that power banks would not be lost. The Figure 9 indicates the image of how the battery will be like charging inside the station.

Figure 7. Power Bank Design

Figure 8. Power Bank Charging System

Figure 9. Power Bank Land and Return Area

IOP Conf. Series: Materials Science and Engineering **658** (2019) 012007 doi:10.1088/1757-899X/658/1/012007

2.3. Secure public Wi-Fi service

The Wi-Fi device will be stored inside the battery exchange station. In Taiwan, the most public Wi-Fi is "iTaiwan". This service now has more than 10,000 hotspots all over mainland Taiwan, even on the high-speed rail (HSR). Therefore, installing this service into the station allows more people to stay connected.

3. Conclusions

As technology improves, it is very hard for people to live a life without assistance from technology. In particular, with the development of the smart phone, power needs have reached the next level. Because electricity demand is so high, power generation is a big problem for every government. Clean energy resources are a better way for people to get electricity. Among all the clean energies, wind is the easiest resource to capture. However, most wind turbines are built by the government, therefore it is sometimes difficult for people to feel the benefits of clean energy. The design of this smart tree allows people to enjoy the benefits of clean energy directly and reduces the use of non-clean generated electricity.

Additionally, a city's convenience is a big factor for visitors. A city that provides complete Wi-Fi for free is another attraction point. Therefore, this design not only uses clean power electricity as the main source, but also employs stations all over the city to provide the secure government-sponsored free Wi-Fi system. This design achieves the goal of go green but smart and gives our next generation a better future.

References

[1] M Wahl 2007 *Designing an H-rotor type Wind Turbine for Operation on Amundsen-Scott South Pole Station* (Stockholm: University of Uppsala/Uptec) p 8

[2] R K Kavade and P M 2017 *Journal of Clean Technologies* **5** pp 353－358

[3] J Miao, W C Hsieh, C C Lai and C S Tai 2013 Journal of Aeronautic, Astronautics and Aviation, Series B 44 pp 31－42

[4] J F Lee 2009 *Wind Energy and Electricity* (Taipei: Coolbooks)

[5] P Tsai 2018 *Public Wi-Fi Risks Worry IT Pros*, https://www.networkcomputing.com/wireless-infrastructure/public-wifi-risks-worry-it-pros/507066099

[6] E Shahin CUJLT *The Hidden Dangers of Public WiFi* **25** pp 205－210

[7] J M Lou,W S Kuo 2018 *The Wind Turbine Concept Design* (Taichung: Feng Chia University)

Towards cloud-based STEP-NC to enhance interoperability in global manufacturing

S Afanasev[1], G Zhao[1, 2] and W Xiao[1, 2]

[1]School of Mechanical Engineering & Automation, 100191 Beihang University, Beijing, China
[2]MIIT Key Laboratory of Aeronautics Intelligent Manufacturing, 100191 Beihang University, Beijing, China

E-mail: xiaowenlei@buaa.edu.cn

Abstract. STEP-NC has been under research for several years, and its manufacturing interoperability is still considered in its infancy. Until recently, seldom STEP-NC compliant CAD/CAM/CNC systems were commercially materialized. The reasons why the CAD/CAM and CNC vendors have not been interested in becoming STEP-compliant are diverse. One of the major is the unmatured interoperability in handling uncertainty in highly dynamic production contexts. STEP-NC requires enormous manufactured data being exchanged among different STEP-compliant systems, whereas little lack of consistency may course the failure of final manufacturing results. The low-volume volatile component production and the complexity of STEP-NC data bring much more risks of uncertainty and unforeseen events, which generally propagate across the different stages of production. The authors believe that the combination of STEP-NC and cloud manufacturing will be the key solution for dealing with the current dilemma of STEP-NC and can promote the intelligent manufacturing into reality. This research proposes a framework of the STEP-NC compliant manufacturing in a cloud environment.

1. Introduction

Modern industry has made an increasing necessity for manufacturing companies to support low-volume volatile component productions and handle the uncertainty from highly dynamic production contexts [1]. Since the inception of STEP-NC, it has been continuously considered as a competitive tool to research and improve different elements of the CAx chain. Thus, it has turned into the main condition for manufacturing businesses to be strategically proactive and competitive [2] on the global scale. In spite of the tremendously encouraging advantages of STEP-NC, its manufacturing interoperability is still relating to a number of issues [2-4]. Comparing to the G-code (ISO 6983) based on CNC programming, STEP-NC (ISO 14649 or AP238) offers much more abundant manufacturing data for the use of object-oriented manufacturing tasks, while it also brings a lot of complexity and difficulties in handling the high-level data models in various CAx systems and realization of the interoperability. The STEP-NC provides the possibility of bidirectional flow of data between CAD/CAM and CNC without any information being lost. It also brings sufficient advancements inside CAD, CAM and CNC systems and the bidirectional data flow of information, which cannot endure any breach or incapability of the CAD-CAM-CNC chain of data [5, 6]. There are numerous hinders in the headway of developing commercialized STEP-compliant CAx systems; the underdeveloped interoperability is nonetheless one of the major issues. Several inconveniences affecting the interoperability of STEP-compliant CAx systems can be presented as follow:

- The STEP-compliant CAD system requires correct understandings of manufacturing features. CAD system requires correct understandings of manufacturing features.
- The STEP-compliant CAM system needs enormous non-geometric data inputs.
- The STEP-compliant CNC system has to transfer the machine tool status upwards to designers.

The introduction of cloud computing into cloud manufacturing gives a reasonable solution to reducing the complexity of handling STEP-NC data and enhancing its interoperability. In cloud-based computing, everything is interpreted as a service (XaaS), e.g. SaaS (Software as a Service), PaaS (Platform as a Service) and IaaS (Infrastructure as a Service) [7]. ISO 14649 or STEP-NC is free of the STEP design data, or technically it was worked out with an objective not to contain the full-fledged design data. This paper presents the STEP-NC based services that facilitate the convenience for both designers and operators.

2. Framework and architecture of global manufacturing cloud

Global Manufacturing Cloud (GMC) is proposed and conceptualized to have the following features throughout the platform. Figure 1 shows the semantic model of GMC. Consumers get access to the GMU via a multi-platform application, browser, computer or mobile device. Multi-platform applications or browsers use a secure Hypertext Transfer Protocol (HTTP) based communication protocol [8].

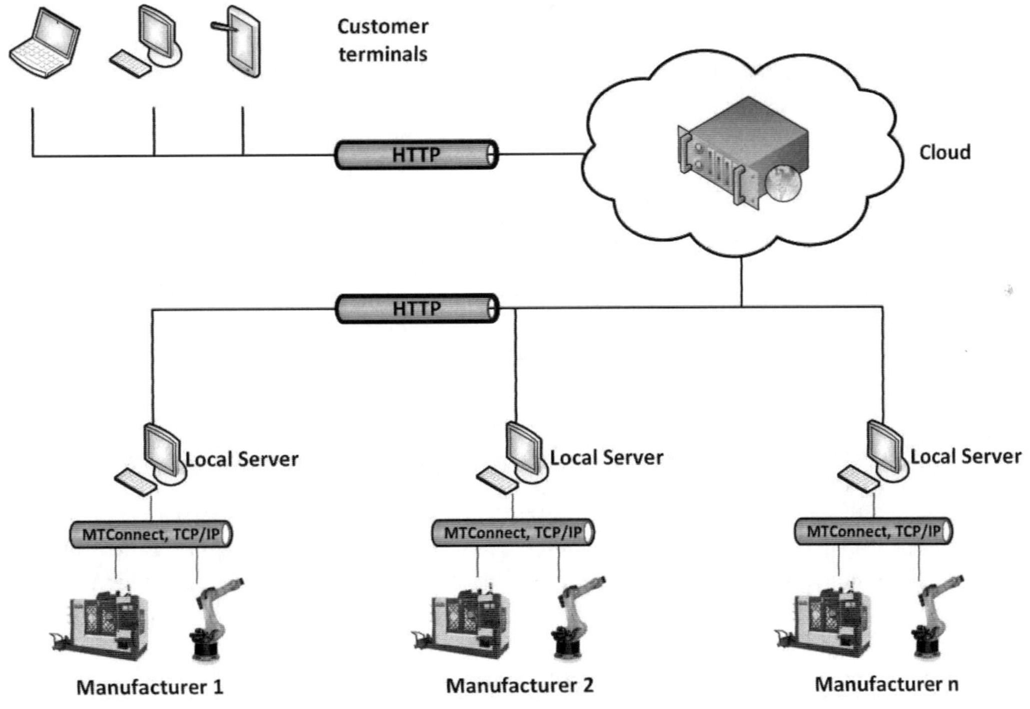

Figure 1. Conceptual model of Global Manufacturing Cloud.

GMC is hosted in the Web-Server. GMC communicates with the manufacturer via the Internet to transmit tasks and to monitor and manage machines. Customers and Administrators connect to the web-server, which then connects to the manufacturer via TCP/IP (Transfer Control Protocol/Internet Protocol), RESTful MTConnect. Each factory has its own local server. The computer is used to manage the machines through controllers through the local network. Local servers send the received data from the cloud and command the controller to perform the manufacturing operations.

IOP Conf. Series: Materials Science and Engineering **658** (2019) 012008 doi:10.1088/1757-899X/658/1/012008

2.1. Framework

In this section is the proposed architecture of GMC. The framework is presented in figure 2. The diagram presents a hierarchical view of communication services and GMC.

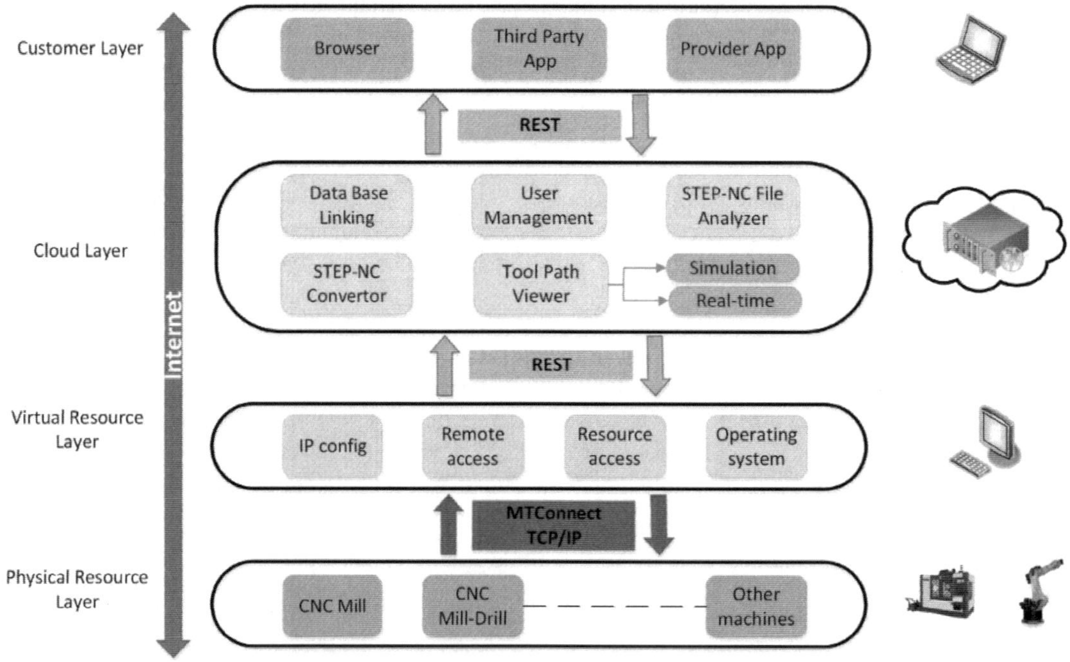

Figure 2. Architecture of Global Cloud Manufacturing Platform.

2.1.1. Physical resource layer This layer includes physical structures of NC machines, motors, drives, Electronic hardware (embedded systems with microprocessors, microcontrollers, memory, input/output), circuits and power supplies. It treats the warehouse facility as a hardware to store software or any future versions of it. Advanced embedded systems of this layer make it possible to have intellectual manufacturing. Manufacturers can publish their resources in the GMC because such resources have Internet access and computational abilities. Manufacturing resources are connected to virtualization resources and can be used for remote monitoring and management throughout the cloud. The TCP/IP Protocol and MTConnect are used to connect manufacturing resources and virtualization resources. MTConnect transfers data for monitoring and collecting data in XML format. TCP/IP communication Protocol is used to send and receive requests.

2.1.2. Virtual resource layer The 'Virtual resource layer' has an advanced operating system, which includes several technologies to identify physical resources (Machines), virtualize and identify them in CMfg service as Infrastructure as a Service (IaaS) [8]. This layer uses resource identification, resource accessing, remote accessing, sharing technologies, along with IoT, and embedded technologies. The virtual resource layer virtualizes the manufacturing resources and represents them in GMC. GMC virtualizes manufacturing resources by publishing accessible RESTful web services across the Internet, refer to figure 2. Manufactures develop and deploy RESTful web services within the servers of the virtual resource layer, which is provided by a local server figure 1. In addition to virtualization, the resource virtualization layer protects production resources by confirming Internet access requests and verifying incoming operations requests from the cloud. This layer uses HTTP-based REST Protocol to interact with the cloud layer.

2.1.3. Cloud service layer The present GMC is not designed to develop any business model but to provide the services primarily for Cloud users instead. The role of the platform administrator is to overlook the entire manufacturing cloud i.e. to integrate and deploy a specific set of functions for the needs of cloud users such as subscription manager of the cloud layer [9] as CMfg services.

For inter-component communication, these cloud services are designed as RESTful web services using HTTP-based REST Protocol. The application layer also interacts with a cloud layer through HTTP based REST protocol. The layer services are presented below.

Subscription Service: This component grants authentication for the user to manage machining tools across the internet when permission is requested. In addition, this service creates a database of user settings and subscriptions.

Cloud Repository Manager stores web virtualization services data and applications. Every announced RESTful virtualized web service has a URL, and these URLs are stored in a database with the required information about the settings. Manufacturers can configure access to web services. Below is the description of published web services.

The Monitoring API Manager provides manufacturers with information how to run published web services from applications. All Internet users have full access to published web services.

The Private API Manager is developed to provide an access to manage and use published web services. For each case publishing manufacturers specify the limits of access. This type of web service provides access for application development and for internal maintenance of production environments.

Security Manager: In order to improve the security of the platform to GMC users provided an authentication token. It is a subscription service component and the security manager will create, monitor and evaluate the token. The authentication token is a time-based and auto-expiring component of the security service.

2.1.4. Application layer Application platforms can be web browsers, mobile applications that are available for download from the repository, built-in applications that allow customers to perform manufacturing operations across the Internet [10].

2.1.5. The communication methods The interaction of the layers is illustrated in figure 2. Cloud components are designed and implemented to provide web RESTful services in a GMC architecture. The architecture is considered from the bottom up. The physical resource layer interacts with the resource virtualization layer using two types of communication such as RESTful MTConnect and TCP/IP. Virtualization layer and the resources layer can interact in two ways: obtain and monitor information status about machine operation and send requests to the machine. Processing commands are transferred over TCP/IP. MTConnect is used to send data to the cloud and store data in XML format. The virtualized resource layer interacts with a cloud layer through RESTful web service. The customer layer interacts with the cloud layer components by making REST calls and receiving an HTTP response. The data is transmitted in XML format.

3. Implementation of GMC

This section discusses the implementation of four web services as follows: STEP-NC File Analyzer and Viewer, Converter STEP-NC, Tool Path Viewer.

3.1. The STEP-NC file analyser

The STEP-NC File Analyzer (SFA) is an accessory program that generates a spreadsheet or CSV files as shown in figure 3. The spreadsheet simplifies the test configuration information at the entity levels and attributes. A customer typically has a hierarchical display of high-level information in the STEP-NC file, where the user can navigate to the individual attributes of the parts. STEP-NC files are used to represent the geometry part, manufacturing information (PMI) and product, which are used for data exchange and interaction between CAx systems. STEP-NC files are also used for storing in a database. The software can also view STEP-NC boundary representation part geometry, tessellated part geometry, PMI for geometric dimensioning and tolerancing (GD&T), supplemental geometry, and finite element models. Other general-purpose STEP file viewers might not display PMI, tessellated part geometry, and finite element models. However, those viewers are usually better at viewing b-rep geometry than the STEP-NC File Analyzer.

Figure 3. The STEP-NC file analyzer of GMC (left), Convertor STEP-NC (right).

3.2. Convertor STEP-NC

STEP-NC convertor is developed using JS, HTML, CSS, C#, EXPRESS, and SQL. It is a utility model of a convertor of the STEP-NC file to XML, to transfer data in XML format to the manufacturing or some part of STEP-NC in XML. This part of STEP-NC designates means by which schemas and data can be performed as an XML document. These schemas are defined by the EXPRESS language and data is carried out by EXPRESS schemas. Key entities which influence the tool path and CNC code would be identified and algorithms are generated considering the key entities as inputs and 'jscad' and '.nc' files as outputs. Figure 3 shows few user interfaces of STEP-NC convertor.

3.3. Tool path viewer

There are two ways to implement the Tool Path Viewer, the first is the OpenJSCAD which offers full control over the modeling process allowing users to write scripts with object-oriented concepts. Another one is NC.js with SignalR, which offers real-time monitoring from manufacturing. The detailed description below.

3.3.1. OpenJSCAD It is a web-based 3D viewer allows the use of JS concepts, libraries, CSG approach for CAD models and interactive parametric modeling with programming guide is identified to be suitable for Tool Path Viewer. Unlike many CAD applications, which offer interactive modeling, it offers full control over the modeling process allowing users to write scripts with object-oriented concepts. Initial testing of tool path templates with interactive parametric modeling is performed against various operations. These toolpath files subsequently refined to incorporate the influence of various other STEP-NC entities and shop floor practices. Algorithms are developed to render simulation for coordinate system, workpiece geometry, feature geometry, tool bit geometry, rapid path, machining path cycles, finishing tool path, return path to retract plane. Working with TPV as discussed in section explains the final output against each of the working steps from STEP-NC. Figure 4 shows the working interface of OJSCAD.

3.3.2. Three.js and SignalR MTConnect, Three.js, and SignalR are three great tools to implement real-time manufacturing monitoring. Real-time web functionality is the facility to provide connected clients with the real-time content using server-side code. Figure 4 shows the real-time movements of the cutting tool and the workpiece.

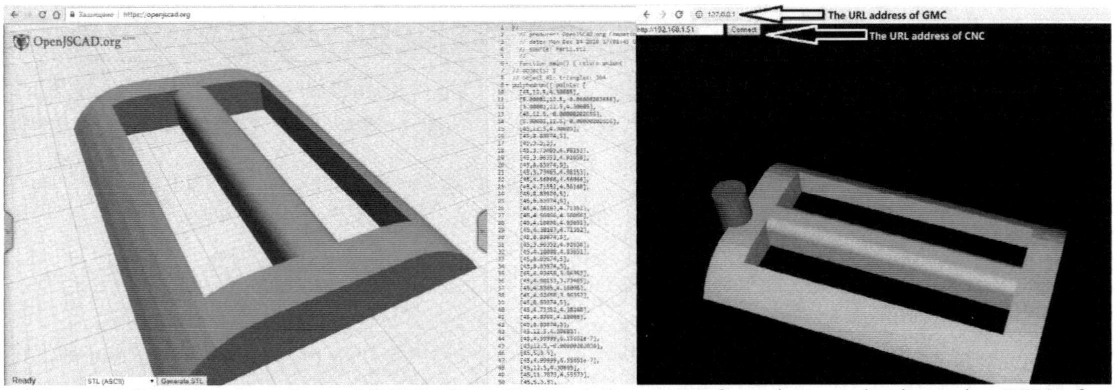

Figure 4. Working interface of OJSCAD (left), Real-time manufacturing monitoring using TPV of GMC (right).

Three.js is the Web interface for the Digital Thread. This provides a rich REST API for process and models as well as a matching client that displays the 3D part models for the machining workpiece, tools, CNC, as well as removal simulation, PMI annotations, MTConnect positional data, QIF face status, and other aspects of a Digital Twin on the Digital Thread.

Three.js allows the creation of Graphical Processing Unit (GPU)-accelerated 3D animations using the JavaScript language as part of a GMC without relying on proprietary browser plugins. This is possible due to the advent of WebGL

SignalR proactively chooses the best accessible transport and takes advantage of multiple transport types. It also takes into account the best available client and server transport. SignalR uses WebSocket, an HTML5 API that provides bi-directional interaction between the browser and the server.

4. Conclusion

The article has presented scalable service-oriented structure, virtualization techniques and methods of communication of the GMC. The international organization for standardization (ICO) has issued the relevant standard - ISO 10303, also called STEP. ISO 14649 Application Reference Model (ARM) to work with CNC is also called STEP-NC. ISO 14649 standard, replaces the use of G-code, facilitating the integration of CAD / CAPP / CAM and CNC systems and the development of MTConnect modules creating the opportunity to monitor the CNC machine in real-time. To enhance connectivity with NC, researchers began to work on various individual STEP-NC systems compliant in the categories of Adapter, Converter, Simulators and CNC controllers. There are very few developed platforms that combine multiple systems. The combination of STEP-NC and CMfg concepts offer a new breed of collaborative and distributable manufacturing solutions. As a fundamental element of the structure, a web server is developed, which link the manufacturer and the customer. It uses the RESTful Internet MTConnect Protocol to acquire the data status of machine tools and monitors their operations over the Internet. The data is exchanged in XML format without converting the STEP-NC file to G-code. Thus, it supports a TCP / IP based solution for the exchange of operational production services in the manufacturing cloud. Users can import and export STEP-NC files from the cloud data associated with their login data implements to work from anywhere. Users simulate the path of the tool in the Tool Path Viewer service. The implemented operational test bench demonstrates the ability to monitor processing activity and the execution of production operations and processes directly from the production cloud across the Internet.

Acknowledgments

Special Program of Ministry of Industry and Information Technology, China.

References

[1] Valente A, Carpanzano E, Nassehi A and Newman S T 2010 *CIRP Annals* **59**(1) pp 441 – 444

IOP Conf. Series: Materials Science and Engineering **658** (2019) 012008 doi:10.1088/1757-899X/658/1/012008

[2] Newman S T, Nassehi A, Xu X, Jr R R, Wang L and Yusof Y 2008 *Robotics and Computer-Integrated Manufacturing* **24**(6) pp 699 — 708
[3] Newman S T 2004 Integrated CAD/CAPP/CAM/CNC manufacture for the 21st century *Proceedings of the14th International Conference on Flexible Automation and Intelligent Manufacturing Conference*
[4] Xu X, Wang H, Mao J, Newman S T, Kramer T R and Proctor F M 2005 CAD/CAPP/CAM/CNC integration *International Journal of Production Research* **43**(17) pp 3703 — 3743
[5] Rauch M, Laguionie R, Hascoet J Y and Suh S H 2012 *Robotics and Computer-Integrated Manufacturing* **28**(3) pp 375 — 384
[6] Xiao W, Zheng L, Huan J and Lei P 2015 *Robotics and Computer-Integrated Manufacturing* **31** pp 1 — 10
[7] Xu X 2012 *Robotics and Computer-Integrated Manufacturing* **28**(1) pp 75 — 86
[8] Tao F, Zhang L, Venkatesh V C, Luo Y and Cheng Y 2011 *Engineering Manufacture* **225** pp 1969 — 1976
[9] Lu Y, Xu X and Xu J 2014 *Journal of Manufacturing Systems* **33** pp 551 — 566
[10] Xi V W and Xun W X 2013 *Robotics and Computer-Integrated Manufacturing* **29** pp 232 — 247

IOP Conf. Series: Materials Science and Engineering **658** (2019) 012009 doi:10.1088/1757-899X/658/1/012009

Novel design for the adjustable knock-out mechanism with a Roberts linkage

C Z Yu[1,2] and F G Wu[1]

[1]Department of Industrial Design, National Cheng Kung University, Taiwan
[2]Department of Animation and Game Design, Toko University, Taiwan

E-mail: ycz@mail.toko.edu.tw

Abstract. This article presents a new adjustable knock-out mechanism of a bolt former with a Roberts linkage and an adjustable crank-rocker linkage. The design requirements of the new mechanism are based on the analysis of an existing adjustable knock-out mechanism. The new adjustable knock-out mechanism mainly consisting of a Roberts linkage, and an adjustable crank-rocker linkage is proposed to meet the specified requirements. The path of the knock-out screw is approximately a straight line and the method presented is illustrated by an example.

1. Introduction

A bolt former is a multistage press for cold forging of bolts [1]. Rod material fed by rollers is cut by a cutter when the required length is reached. It is then delivered to the first forging mold for the molding. Several stages have to be gone through before rod material could be gradually formed into the required shape. The shaping bolt and the forging mold are stuck together in each stage of molding, thus it is necessary to use the knock-out pin to knock the shaped bolt out of the mold after the molding process in each stage. In order to knock out bolts of different lengths, an adjustable knock-out mechanism is required to satisfy the need of variable knock-out strokes.

The concept design of adjustable knock-out mechanisms for the bolt former presented in the patent can be classified into two types: (a) adjustable linkages [2-4], and (b) adjustable cam-follower mechanisms [5-7]. An adjustable slider-crank linkage with an adjustable crank [2] is used to generate various knock-out strokes in bolt formers, but the extreme positions of the slider in various knock-out strokes are not the same. An adjustable four-bar linkage [3-4] is used in bolt formers to alter the knock-out stroke. The knock-out pin is driven by the output link of a four-bar linkage. The input link of the four-bar linkage is controlled by a cam. The link length of the input link can be adjusted to change the angle of the output link and knock-out stroke. Wang [5] proposed a knock-out cam-follower mechanism consisting of a knock-out pin, a swing cam, and an oscillating follower with a curved surface. The follower also serves as a pivoted lever arm for pushing the knock-out pin. The curved surface on the pivoted lever arm can be adjusted by two screws to generate various knock-out strokes. Tan [6] presented a knock-out cam-follower mechanism in which the follower of the mechanism includes two rollers; the cam can be adjusted to contact one roller or another in order to generate different stroke lengths. Yoshiichi [7] provided a knock-out cam-follower mechanism consisting of a cam-follower mechanism and a knock-out pin. The cam drives the oscillating roller follower to push the knock-out pin. The cam profile can be adjusted by an adjustment screw to generate various knock-out strokes to manufacture bolts of different lengths. Based on the concept stated above, a method is needed to design the cam profile with adjustment.

There are several approaches to design the cam profile without adjustment [8-14]. Hwang and Yu [15] presented an optimization method to design the cam profile with adjustment. For the cam profile with a sixth-degree polynomial segment, besides the six specified boundary conditions, a variable condition is used as a design variable of an optimization design. The kinematic characteristics of the new knock-out cam-follower mechanism synthesized are better than those of the existing mechanism.

The existing knock-out mechanisms [2-7] generally have two disadvantages: (1) The path of the knock-out screw is an arc, but the output of the knock-out mechanism required is a linear reciprocating movement for the knock-out pin; (2) The knock-out screw strikes the knock-out pin to make the noise. This article presents a new concept of adjustable knock-out mechanism with a Roberts linkage and an adjustable crank-rocker linkage. The path of the knock-out screw is approximately a straight line.

There are several approaches for the synthesis of linkage. Ahmad and Waldron [16] used an adjustable fixed pivot of the driven link to design adjustable four-bar linkages for motion generation. Krishnamoorthy and Kothadiya [17] presented a method to synthesize a four-bar linkage with an adjustable fixed pivot to obtain two intersecting straight lines. Wang and Sodhi [18] adjusted the moving pivot position of a four-bar linkage to generate multi-phase motion. Zhou and Ting [19] adjusted the position of the slider guider of a slider-crank linkage for multiple path generation. Wu and Chen [20] presented an adjustable four-bar linkage with an adjustable length link to synthesize linkages passing through more precision points. Zhou [21] adjusted the length of the coupler or the driven link for precise function generation. Pennock and Israr [22] presented an adjustable six-bar linkage with an adjustable input crank to synthesize the extreme positions of the output link and coupler link. An adjustable crank-rocker linkage is used to generate various oscillating angle of rocker in the new concept.

The purpose of this article is to propose a new adjustable knock-out mechanism in which the Roberts linkage generates straight knock-out paths and an adjustable crank-rocker linkage generates various strokes.

2. Design requirements of the knock-out mechanism

The knock-out mechanism of a bolt former used in a local company in Taiwan is shown in Fig. 1. It mainly consists of a crank-rocker linkage, an adjustable cam-follower mechanism, and a knock-out pin. The knock-out pin is struck forward by a knock-out screw on a pivoted lever arm. The pivoted lever arm is an oscillating roller follower controlled by a swing cam that is driven by the crank-rocker linkage. The cam profile can be adjusted around the adjustment axis to alter the starting position of the follower in order to generate various strokes of the knock-out pin. According to the kinematic analysis of the existing adjustable knock-out mechanism, the maximum strike velocity between the knock-out screw and knock-out pin is 1248.1mm/s when the knock-out stroke is 81.4mm. The horizontal component of the position of the knock-out screw for various knock-out strokes varied continuously from 81.4mm to 226.4mm [15].

According to the kinematic analysis of the existing adjustable knock-out mechanism, the design requirements for the new adjustable knock-out mechanisms are specified as follows:

(1) The horizontal component of the position of the knock-out screw for various knock-out strokes varied continuously from 81.4mm to 226.4mm (Fig. 2).

(2) All the extreme positions of the knock-out pin for different knock-out strokes are the same because the tongs catch different length bolts at the same location.

(3) The knock-out mechanism is driven by a DC motor.

(4) The output of the knock-out mechanism is a linear reciprocating movement of the knock-out pin.

(5) The motion of the knock-out pin must be controlled by a mechanism.

(6) The knock-out screw doesn't strike the knock-out pin to eliminate the noise.

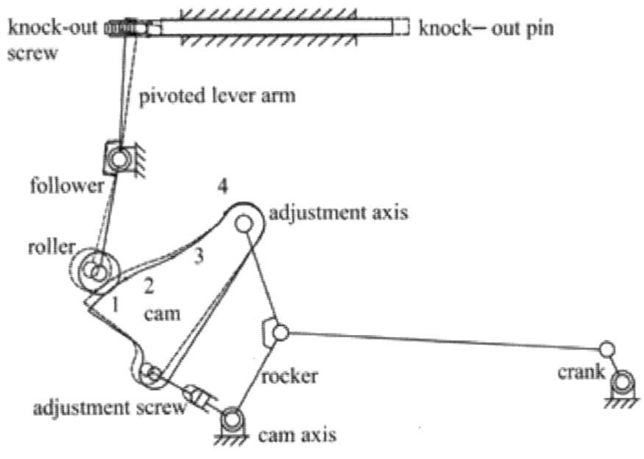

Figure 1. Adjustable knock-out mechanism [15].

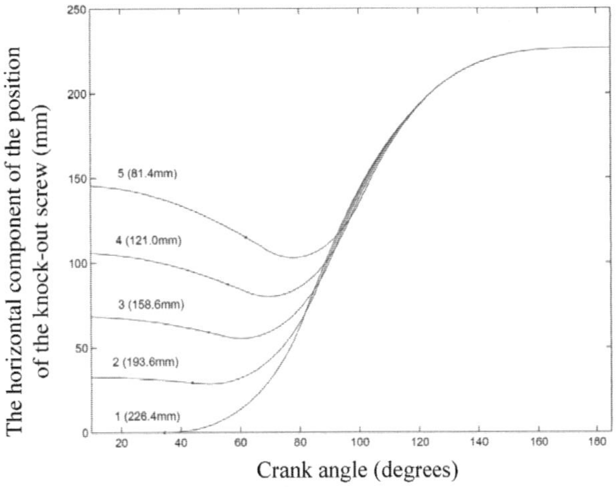

Figure 2. The horizontal component of the position of the knock-out screw for various knock-out strokes of the previous design [15].

3. A new adjustable knock-out mechanism

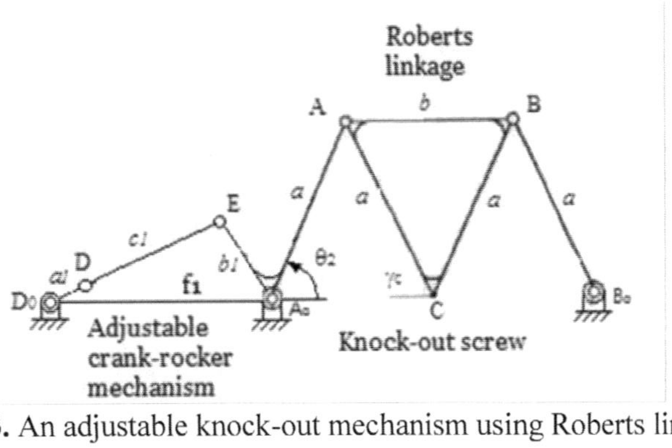

Figure 3. An adjustable knock-out mechanism using Roberts linkage.

Based on the functional requirements stated above, a new adjustable knock-out mechanism is presented below. The Roberts linkage must guide the knock-out screw to generate a straight-line path so as to satisfy the requirement for the linear reciprocating movement of the knock-out pin. Linkages generating a straight-line path are named straight-line linkages. Many popular straight-line linkages have been widely used for a long time [23]. The Roberts linkage is selected herein for the knock-out mechanism. The Roberts linkage is driven by the rocker of a crank-rocker linkage D_0DEA_0. Therefore, the oscillating angle of link A0A is the same as that of rocker A_0E. The Roberts linkage, with the dimensions that $A_0A=B_0B=AC=BC=a$, AB=b, and $A_0B_0=2b$, generates a coupler curve γ_c as shown in Fig. 3. The straight-line portion of the coupler curve traced by the knock-out screw attached to the coupler point C is suitable for guiding the knock-out pin. Because the final positions of the knock-out screw on the coupler curve for different knock-out strokes should be the same, the length of link D_0D and DE in the limiting position is adjustable by shifting pivot D along line D_0E. The new adjustable knock-out mechanism proposed above consists of a Roberts linkage and an adjustable crank-rocker linkage for knocking out bolts of different lengths by adjusting the lengths of link D_0D and link DE.

4. Synthesis procedure

In order to satisfy the kinematic requirements stated above, a procedure for synthesizing the dimensions of the new adjustable knock-out mechanism is presented as follows:

(1) Specify the design demands of the knock-out mechanism, e.g., the required maximum and minimum knock-out strokes, S_{max} and S_{min}, the working range $\Delta\phi$ of the crank D_0D, and the starting angle ϕ_0 of the crank D_0D.

(2) Construct the mathematical model of the position deviation of coupler point C from the desired straight line as follows: Adjust the length of link DE from c_1 to $c_1+\Delta d$; the length of link D_0D is thus $a_1-\Delta d$. The adjusted linkage will generate a smaller oscillating range. The coordinates of coupler point C can be expressed as follows:

$$x_C(\theta_2) = a\cos\theta_2 + a\cos\left[\theta_3 - \cos^{-1}\left(\frac{b}{2a}\right)\right] \tag{1}$$

$$y_C(\theta_2) = a\sin\theta_2 + a\sin\left[\theta_3 - \cos^{-1}\left(\frac{b}{2a}\right)\right] \tag{2}$$

Where

θ_2 is the angular positions of link A_0A

θ_3 is the angular positions of link AB

$$\begin{aligned}
&\theta_3 = 2\tan^{-1}\left[\frac{-E+\sqrt{E^2-4DF}}{2D}\right]\\
&D = \cos\theta_2 - k_1 + k_2\cos\theta_2 + k_3\\
&E = -2\sin\theta_2\\
&F = k_1 + (k_2-1)\cos\theta_2 + k_3\\
&k_1 = 2b/a\\
&k_2 = 2\\
&k_3 = -(a^2+4b^2)/2ab
\end{aligned} \tag{3}$$

The knock-out stroke S is the difference of the x coordinates of coupler point C between the starting and final positions, and can be written as:

$$S = x_c(\theta_{2f}) - x_c(\theta_{2s}) = b - a\cos\theta_{2s} - a\cos\left[\theta_{3s} - \cos^{-1}\left(\frac{b}{2a}\right)\right] \tag{4}$$

Where θ_{2s} and θ_{2f} are the starting and final angular positions of link A_0A, respectively. θ_{3s} is the starting angular position of link AB.

(3) For a specific knock-out stroke S_{max}, design the link lengths of an Roberts linkage and the total oscillating range $\Delta\Psi_{max}$ of rocker A_0E.

(4) For a specific knock-out stroke S_{min}, design the total oscillating range $\Delta\Psi_{min}$ of rocker A_0E.

(5) According to the total oscillating angle $\Delta\Psi_{max}$ of the rocker A_0E, the working angle $\Delta\phi$ of the crank D_0D, and the starting angle ϕ_0 of the crank D_0D, synthesize the link lengths of the adjustable crank-rocker linkage which controls the motion of the Roberts linkage. The link lengths can be written as [24]:

$$a_1 = \left[\frac{-\sin \Delta\psi_{max}/2}{\sin \dfrac{\Delta\phi - \Delta\psi_{max}}{2}} \cos\left(\phi_0 + \Delta\phi/2\right) \right] f_{11} \tag{5}$$

$$b_1 = \left[1 + \left(\frac{a_1 + c_1}{f_1}\right)^2 - 2\left(\frac{a_1 + c_1}{f_1}\right)\cos\phi_0 \right]^{\frac{1}{2}} f_1 \tag{6}$$

$$c_1 = \left[\frac{\sin(\phi_0 + \Delta\phi - \Delta\psi_{max}) - \sin(\phi_0 + \Delta\phi)}{\sin(\Delta\phi - \Delta\psi_{max})} - \frac{a_1}{f_1} \right] f \tag{7}$$

where a_1, b_1, c_1, and f_1 are the lengths of the crank D_0D, rocker A_0E, coupler link DE, and fixed link D_0A_0, respectively.

(6) According to the total oscillating angle $\Delta\Psi_{min}$ of the rocker A_0E and the lengths of the crank D_0D, rocker A_0E, coupler link DE, and fixed link D_0A_0, synthesize the adjustable value Δd of crank D_0D [24]:

$$\Delta\psi_{min} = \cos^{-1}\frac{b_1^2 + f_1^2 - (a_1 + c_1)^2}{2b_1 f_1} - \cos^{-1}\frac{b_1^2 + f_1^2 - (c_1 - a_1 + 2\Delta d)^2}{2b_1 f_1} \tag{8}$$

Example: Design a Roberts knock-out mechanism shown in Fig. 2 to satisfy the following design requirements: (a) the maximum knock-out stroke is 226.4mm, (b) the minimum knock-out stroke is 81.4mm, (c) the working range of the crank D_0D is 150.4°, and (d) the starting angle of the crank D_0D is 55.6°.

(1) For a specific knock-out stroke S_{max} is 226.4mm, design the link lengths of an Roberts linkage and the total oscillating range $\Delta\Psi_{max}$ of rocker A_0E is 16.6°. The link lengths a and b are 500mm and 500mm, respectively.

(2) For a specific knock-out stroke S_{min} is 81.4mm, design the total oscillating range $\Delta\Psi_{min}$ of rocker A_0E is 5.5°.

(3) Substituting $\Delta\Psi_{max}$=16.6°, $\Delta\phi$=150.4°, and ϕ_0=55.6° in Eqs. (5)~(7), the link lengths of the crank D_0D, rocker A_0E, coupler DE, and fixed link D_0A_0 are 102.5mm, 845.4mm, 278.6mm, and 1000mm, respectively.

(4) Substituting a_1=102.5mm, b_1=845.4mm, c_1=278.6mm, f_1=1000mm, and $\Delta\Psi_{min}$=5.5° in Eqs. (8), the adjustable value Δd of the crank D_0D and coupler DE is 63.8mm, The value of Δd is from 0mm to 63.8mm, corresponding to a knock-out stroke from 226.4mm to 81.4mm.

5. Conclusions

This paper presents a method for the configuration design and dimensional synthesis of an adjustable knock-out mechanism of a bolt former with a Roberts linkage. This adjustable knock-out mechanism has the following advantages:

(1) This article presents a creative concept of adjustable knock-out mechanism with a Roberts linkage. The path of the knock-out screw is approximately a straight line.

(2) A new adjustable knock-out mechanism is used to eliminate the noise, because the knock-out screw doesn't strike the knock-out pin.

References

[1] Mcbain N S and Uhic S J 1982 *Choice of technique in bolt and nut manufacture* (Edinburgh:

Scottish Academic Press)
[2] Waterbury Farrel Foundry Co 1955 *UK Pat.* 726737
[3] Flamme W 1981 *EU Pat.* 0041633
[4] Helmut D 1987 *US Pat.* 4635463
[5] Wang S Y 1989 *R. O. C. Pat.* 46312
[6] Tan S Y 1996 *R. O. C. Pat.* 118846
[7] Yoshiichi S 2002 *Jpn. Pat.* 2002361357
[8] Chen F Y 1982 *Mechanics and Design of Cam Mechanism* (New York: Pergamon Press)
[9] Wilson C E and Sadler J P 1993 *Kinematics and Dynamics of Machinery* (New York: Harper Collins College Publishers)
[10] Tsay D M and Wei H M 1995 *ASME J. Mechanical Design* **118** pp 259－265
[11] Shin J H et al 1996 *Proc. of the 6th Symp. on Transport Phenomena and Dynamics of Rotating Machinery* **1** pp 178－186
[12] Shin J H et al 2000 *J. of the Korean Society of Precision Engineering* **17**(2) pp 185－192
[13] Shin J H and Zhang Y H 2004 *ASME J. Mechanical Design* **126** pp 183－188
[14] Biswas A et al 2004 *Mech. Mach. Theory* **39** pp 645－656
[15] Hwang W M and Yu C Z 2005 *Proc. of the IMechE Part C : Journal of Mechanical Engineering Science* **219** pp 767－774
[16] Ahmad A and Waldron K J 1979 *Mech. Mach. Theory* **143** pp 405－411
[17] Krishnamoorthy S and Kothadiya A J 1987 *Mech. Mach. Theory* **222** pp 107－114
[18] Wang S J and Sodhi R S 1996 *Mech. Mach. Theory* **314** pp 459－474
[19] Zhou H and Ting K L 2002 *Mech. Mach. Theory* **37** pp 499－509
[20] Wu T M and Chen C K 2005 *Appl. Math. Comput.* **160** pp 309－316
[21] Zhou H 2007 *Proc. of the 31st Mechanisms and Robotics Conf.* **8** pp 503－509
[22] Penock G R and Israr A 2009 *Mech. Mach. Theory* **442** pp 306－323
[23] Norton R L 1989 *Design of machinery* (New York: McGraw-Hill Book Company)
[24] Chiang C H 2000 *Kinematics and design of planar mechanisms* (Krieger Publishing Company)

IOP Conf. Series: Materials Science and Engineering **658** (2019) 012010 doi:10.1088/1757-899X/658/1/012010

The determinants of smart tourism behavior: evidence in wine tourism context

TsaiFaYen[1], YuJanShen[2]and ChienJen Hung[2]

[1]School of Economics, Sichuan UniversityofScienceandEngineering, China
[2]Fuzhou Universityof International Studiesand Trade, China

E-mail: 1722997311@qq.com

Abstract. The main purpose of this study is to investigate theinfluence of satisfaction, involvement, and subjective norms on tourists 'behavioral intentions at a historic district at Yibin, China. This study also attempts to examine the mediating effect of involvement, and subjective norms on the paths from satisfaction to intentions. Data were collected via a nonsite survey and analyzed using a structural equation modeling approach. Results indicate that satisfaction, involvement, andsubjective norms directly contribute to the formation of tourists' intentions of the historic district. Tourists' involvement is positively associated with tourists' satisfaction. Both involvement and subjective norms fully mediates the relationship between satisfaction and intentions. Theoretical and managerial implications are elaborated.

1. Introduction

Smart tourism has become more and more popular in mainland China. Different types of events and activities have been classified as smart tourism, leading to the misuse of the term[1]. Smart tourism is viewed as a social phenomenon arising from the convergence of citizens‘ daily life in business and tourism ecosystems[2]. The concept of smart tourism comes from smart city and tourism systems, which involve flows and exchanges of data and information for satisfying tourists‘ needs. Such flows and exchanges constantly depend on interdependence and interact with external systems (e.g., transportation, healthcare, and payment systems), along with resource sharing for optimization. Optimization is achieved through advance technologies in the ecosystems with the support of residential government policies and systems integration [3, 4].The word —Smart‖ implies —smartness,‖ which means not only fast, convenient, cheap, and intelligent for a traveler, but also efficient, effective, productive, and creative for business in terms of providing and consuming tourism products and services through a network of cooperating businesses [2]. Studies related to smart tourism in past five years have focus on the technology and few studies focus on its application. Given this lack of clarity, practices such as putting previous wine into new bottles and rushing headlong into mass action will simply lead to a new round ofinefficient resource allocation and consequent lowered confidence on the part of enterprises and government authorities [1]. Thus, this study aims at investigating its application on wine tourism. The purpose of this study is1) to verify the effects of involvement, subjective norm, and satisfaction on behavioral intentions; 2) assessing the moderated effects of involvement and subjective norm on satisfaction-behavioralintentions relation; and 3) drawing some implications. Research framework of this study is presented in Figure 1.

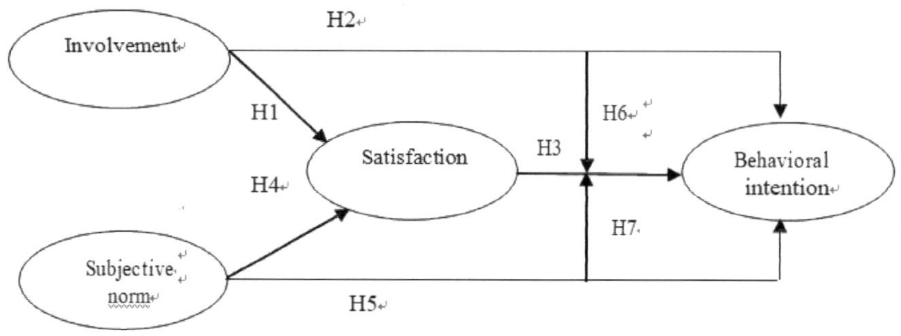

Figure 1. Researchframework

2. Literature review and conceptual model

2.1. The concept of involvement and its impact on satisfaction and behavioral intentions

Involvement has been conceptualized as ―a person's perceived relevance of the object or product based on inherent needs, values, and interests‖ [9]. The relationship between involvement and satisfaction of a tourist is under debate in the tourism literature. While [5] report a positive relationship between involvement and satisfaction in Litchi Bay, China; Prayag and Ryan[6]fail to support this relationship in their study. Others identify involvement as an antecedent of tourist satisfaction and future visit intentions. For example [7, 8] conclude that tourists who are highly involved in the activities during a festival/tour are more satisfied with their experience and express stronger intentions to return compared to individuals who are less involved. Tourists' involvement in this study is mainly evaluated by their participationindestination activities. In this vein, itis postulated that:

H1: High levels of involvement positively affect tourists' satisfaction with a wine tourism site.

Several studies provideempiricalevidence for the relationships between involvement and behavioral intention. Studies on food involvement have suggested that involved consumers make healthier food choices and that further research is necessary on the role of involvement in consumer behavior [9, 10]. Involvement as an antecedent to outcomes such as willingness to purchase wine [11] and food involvement as a direct predictor of loyalty[12] suggest that involvement can have a direct relationship with behavioral outcomes [13]. Given that visitors have high involvement with a wine tourism site, it is believe that this will lead to positive word of mouth, recommendation, and future visitation. Thus, high levels of involvement positively affect their future intention.

H2: High levels of involvement positively affect tourists' behavioral intention with a wine tourism site.

2.2. The concept of satisfaction and its impact on behavioral intentions

Satisfactionrefers to a consumer-perceived discrepancy between expectation andperformance before and after consumption [14]. Satisfaction is an important concept in tourism[6]. A number of studies show that understanding tourist satisfaction is essential to successful destination marketing because it not only impacts the consumption of goods and services, but also the future selection of a destination and intention to visit by tourists [15-18]. Customers often make a decision to purchase/repurchase after evaluating whether their experiences with a product/service are satisfactory/pleasurable [19-23]. Assuring customer satisfaction can therefore be viewed as the capability of companies to meet customers‘ expectations, with a high degree of satisfaction being a source of companies‘ competitive advantages [14-24]. In their empirical research examining tourists‘ behaviors, [25] empirically demonstrated that tourists‘ satisfaction formed based on enjoyment and economic benefits, and directly and indirectly engenders intention to revisit in P2P accommodation. Similarly, [23] clearly showed that consumers‘ satisfaction level plays a significant role in their intention to revisit the restaurant and other restaurants of the same hotel formation. When tourists feel satisfied, their favorable intentionsforaproduct/service and their desire toward purchasing and experiencing it generally increase [23]. Given that visitors have high satisfaction with a wine tourism site, it is believe that this will lead to positive word of mouth, recommendation, and future visitation. Thus, high levels of satisfaction positively affect their future intention.

H3: High levels of satisfaction positively affect tourists' behavioral intention with a wine tourism site.

2.3. The concept of subjective norms and its impact on satisfaction and behavioral intentions

According to [26], subjective norms refer toperceived pressures on a person to perform a given behavior and the person's motivation to comply with those pressures.'' Thus, subjective norms reflect how the customer is affected by the perception of some significant referents (e.g., family, friends, and colleagues, among others) of his/her behavior [27]. When a consumer is affected by the perception of some significant referents, s/he is likely to present satisfied status and positive behavioral intention. [28] show that subjective norm positively influence customer's satisfaction in mobile shopping. Similarly, [29] find that subjective norms playasignificant role in predicting satisfaction and behavioral intention in O2O service. Given that visitors have high subjective norms with a wine tourism site, s/he is likely to be satisfactory and reveal positive word of mouth, recommendation, and future visitation. Thus, high levels of subjective norm are likely to affect their satisfaction and behavioral intention.

H4:High levels of subjective norm positively affect tourists' satisfaction with a wine tourism site.

H5:High levels of subjective norm positively affect tourists' behavioral intention with a wine tourism site.

2.4. The moderated effect of involvement on satisfaction-behavioral intention relation

Within many studies of marketing research on consumer behavior, it is noted that involvement often exhibits a moderating effect on proposed relationships that include behavior outcomes such as satisfaction, intentions, and loyalty [30, 31]. [32] argued that there are two ways in which tourists can become involved in the co-creation process: passively or actively[33]. Passive involvement refers to interactions that are generally controlledbythe setting—the tourist provides some input to the overall experience such as visiting a theme park. On the other hand, active involvement allows the tourist to immersethemselves in the experience, taking on responsibility for every step in the process [18]. [34] argued that causation between customer satisfaction and loyalty was moderated by their level of involvement, and [35] verified that customer level of involvement had a positive moderating effect on causation between satisfaction, intent of repurchase, and loyalty. Given that visitors have high involvement with a wine tourism site, s/he is likely to immerse themselves in the experience of wine tourism site. S/he is likely to have better word of mouth, recommendation, and future visitation even s/he is not satisfied with the wine tour. Thus, the relationship between satisfaction and behavioral intention to the wine tourism site is likely to be moderated by the tourist's level of involvement.

H6: The relationship between satisfaction and behavioral intention to the wine tourism site is moderated by the tourist's level of involvement.

2.5. The moderated effect of subjective norms on satisfaction -behavioral intention relation

Wine tourism has been recognized as a form of theme tourism, comprising visits to vineyards, wineries, wine festivals, and wine shows in wine tourism destinations [36]. Visitors can taste a variety of wine and they often buy some wine as souvenirs after traveling [37]. Therefore, some reference group of visitor will feel a little confusion about taking a trip to wine tourism site. They might have high probabilities to stress visitor not to travel to wine tourism site even visitor is satisfied with the wine tour and want to revisit again. On the other hand, visitor with high stress from reference group is likely to have high subjective norm and consequently the relationship between satisfaction and behavioral intention to the wine tourism site is likely to be moderated by the high level of tourist's subjective norm. [37] reports that subjective norm can destroy the causal relation between destination image and word of mouth. Similarity, Lin, Liu, and [37] state that subjective norm can decline the causal relation between satisfaction and word of mouth in logging industry. Thus, the following hypothesis is proposed:

H7: The relationship between satisfaction and behavioral intention to the wine tourism site is moderated by the visitor's level of subjective norm.

3. Method

3.1. Measurement tools

The measurement instruments for data collection were adopted from existing multiple-item scales validated by previous studies [5, 10, 27, 38-40]. The adopted measures were altered to be adequate in the present research setting. In particular, involvement was measured with three items (e.g., I like being involved in making visits of wine sites‖; I attach great importance to visiting wine sites‖; —Purchases decisions for wine tourism are very important to me‖). Subjective norm was assessed with three items (e.g., People whose opinions I appreciate think that it is okay for me to visit this wine site‖; Most people I valued think that I should visit this wine site‖; —It is expected that I visit this wine site to see some wine products/ services‖;). Satisfaction was measured with three items (e.g., I am happy with the products/services I have visited on this wine site‖; I am generally happy with having visited from this wine site‖; In general, I have liked visiting this wine site‖). Moreover, behavioral intention was measured with three items (e.g., I say positive things about the experience at wine site‖; I will recommend visiting wine site to anybody who seeks my advice‖; —I will revisit the wine site in the future‖).

A five-point Likert type scale, which anchored from strongly disagree (1) to strongly agree (5), was used for all of the questions. The respondents were also requested to provide their socio-demographic information. The surveyquestionnaire including the measures described above and demographic questions were pre-tested with 30 undergraduate and three faculty members whose frequency of wine tourism visit is relatively high and improved accordingly. The questionnaire was furthermore reviewed and perfected by academic experts whose major is tourism/hospitality management and industry experts who are working in a wine tourism site.

3.2. Data collection and demographic characteristics

In order to collect data, a personal interview survey methodology was utilized. This face to face field survey approach allowed us to reach a wider sampling range and include a greater number of wine visitors during the survey period. All eligible survey respondents were instructed to read the study description carefully when accessing the survey, and were requested to fill out the questionnaire. The participants were also requested to indicate the name of the wine site in which they visited. A total of 650 completed questionnaires were gathered through these processes. After the screening process by removing unusable responses and extreme cases, 630 usable responses were retained for data analysis. The socio-demographic profile of the survey respondents was examined. Of 630 participants, 46.5% are males and 53.5% females, most of them between 20 and 39 years old (41.2%), with secondary education (35.2%).

3.3. Data analysis

The collected data through the previously described data collection procedure was analyzed with SPSS and AMOS Version 20. A two-step process was employed [41]. In particular, a measurement model was initially evaluated by conducting confirmatory factor analysis (CFA). Subsequently, a structural equation modeling (SEM) was estimated.

4. Results

4.1. Measurement model evaluation

A measurement model comprising all of the construct measures and nine latent variables was generated via the CFA with a maximum likelihood estimation method. The goodness-of-fit statistics of the CFA revealed an adequate fit to the data ($\chi2$ = 215, df = 43, $\chi2$/df = 5, p < 0.001, GFI= .945; AGFI= .901; CFI= .953; RMSEA= .080). All items were significantly loaded to their associated latent construct (p < 0.01). Table 1 presents the details of the CFA results. As reported in Table 1, the average variance extracted (AVE) of the constructs were greater than the 0.50 threshold suggested byHair et al. (1988). The values ranged from 0.548 to 0.695. Thus, convergent validity for each construct was supported (Fornell and Larcker, 1981). In addition, these AVE values exceeded the square of correlations between variables (Table 2). Hence, discriminant validity was evident (Fornell and Larcker, 1981). Composite reliability was then calculated. Our calculation revealed that the values were equal to or higher than 0.76, confirming the adequate level of internal consistency among the items for each latent variable.

Table1. Measurementmodelassessment

Construct	Items	λ	t-values	SMC	CR	AVE
INV	INV1	.807	-	.651	.872	.695
	INV2	.870	22.981	.756		
	INV3	.822	22.084	.675		
SN	SN1	.670	-	.449	.836	.633
	SN2	.904	17.210	.818		
	SN3	.796	17.427	.633		
SA	SA1	.649	-	.421	.783	.548
	SA2	.773	15.213	.597		
	SA3	.790	15.390	.625		
BI	BI1	.739	-	.546	.832	.626
	BI2	.911	20.182	.830		
	BI3	.709	17.177	.502		
Model fit	χ^2=215; df= 43(p=.000); χ^2/df= 5; GFI= .945; AGFI= .901; CFI= .953; RMSEA= .080					

INV: involvement; SN: subjective norm; SA: satisfaction; BI: behavioral intention; λ:standardized factor loading; SMC: square multiple correlation; CR: composite reliability; AVE: average variance extracted

Table2. Correlation and discriminant validity

Items	M	SD	INV	SN	SA	BI
INV	10.11	2.89	**.695**			
SN	11.10	2.65	.150**	**.633**		
SA	11.31	2.35	.334**	.280**	**.548**	
BI	12.03	2.36	.368**	.593**	.273**	**.626**

**p<.05 (two-tailed); Diagonal elements were average variance extracted and off-diagonal were the coefficient of the correlation.
M:mean ;SD: standard deviation; INV: involvement; SN: subjective norm; SA: satisfaction; BI: behavioral intention

4.2. Structural model evaluation
Following the evaluation of the measurement structure, the SEM was conducted using a maximum likelihood estimation approach in order to validate the proposed theoretical framework and to test the hypothesized relationships among study constructs. Three models were adopted for testing the hypotheses. INV, SA, SN, and BI were conducted in the first model. Further the interaction item INV*SA was added in model 2 and finally the second interaction item SN*SA was added in model 3. As indicated in Table 3, the goodness-of-fit statistics showed the acceptable fit to the data in model 1(χ^2 = 215, df = 43, χ^2/df = 5, p < 0.001, GFI= .945; AGFI= .901; CFI= .953; RMSEA= .080).

Overall, the hypotheses in model 1 were supported except H5. Satisfaction was well accounted for by its predictors (R^2 = 0.226). Behavioral intention was also well accounted for by its predictors (R^2 = 0.516).

Furthermore, the path INV*SA was added and the goodness-of-fit statistics showed the acceptable fit to the data in model 2(χ^2 = 243, df = 51, χ^2/df = 4.77, p < 0.001, GFI= .943; AGFI= .899; CFI= .949; RMSEA= .077). Again, all the hypotheses in model2 were supported except H5. Satisfaction was well accounted for by its predictors (R^2 = 0.226). Behavioral intention was also well accounted for by its predictors (R^2 = 0.544). In particular, that the relationship between satisfaction and behavioral intention to the wine tourism site is moderated by the tourist's level of involvement (H6) was evident. Moreover, the path SN*SA was added and the goodness-of-fit statistics showed the acceptable fit to the data in model 3(χ^2 = 283, df = 60, χ^2/df = 4.70, p < 0.001, GFI= .940; AGFI= .894; CFI= .942;RMSEA= .077). All the hypotheses in model 3 were supported except H5. Satisfaction was well accounted for by its predictors (R^2= 0.229). Behavioral intention was also well accounted for by itspredictors (R^2 = 0.541). In particular, that the relationship between satisfaction and behavioral intention to the wine tourism site is moderated by the tourist's level of involvement (H6) was confirmed. The results of the SEM are exhibited in details in Table 3.

Table3. Structural model assessment

Paths	M1		M2		M3	
	β	t	β	t	β	t
H1:INV-SA	.343***	7.095	.342***	7.089	.344***	7.123
H2:INV-BI	.173***	4.096	.191***	4.581	.192***	4.585
H3:SA-BI	.624***	10.465	.573***	9.775	.573***	9.820
H4:SN-SA	.262***	5.363	.263***	5.367	.265***	5.423
H5:SN-BI	.018	.454	.020	.523	.021	.525
H6:INV*SA-BI			-.161***	-4.455	-.167***	-4.545
H7:SN*SA-BI					.004	.126
R^2_{SA}	.226		.226		.229	
R^2_{BI}	.516		.544		.541	
Model Fit						
χ^2	215		243		283	
df(p)	43(.000)		51(.000)		60(.000)	
χ^2/df	5		4.77		4.70	
GFI	.945		.943		.940	
AGFI	.901		.899		.894	
CFI	.953		.949		.942	
RMSEA	.080		.077		.077	

*p<.05; **p<.01; ***p<.001
INV: involvement; SN: subjective norm; SA: satisfaction; BI: behavioral intention

4.3. Discussion

This study tested a model that proposed relationships among four constructs: involvement, subjective norm, satisfaction, and behavioral intention. The findings of this study indicate that involvements can contribution to satisfaction. Given the nature of our screening questions and the way the construct of involvements was worded, the study created a setting in which the respondent had the opportunity to engage meaningfully in the process of wine tourism. This finding is essentially consistent with past research [8, 42]. Moreover, the study confirms past research that demonstrates the link between involvement and behavioral intention [10, 11, 13]. Involvement helps improve visitors' behavioral intention during the trip and after they return home. Therefore, the study argued that the findings of this study contribute to the growing body of knowledge in understanding the wine tourism by establishing a theory-based empirical link between involvement and behavioral intention. This link had not been empirically demonstrated, as until this study, no study had focused on the aftereffects of participation in wine tourism when talking about smart tourism.

Furthermore, in line with previous studies [19, 20, 22, 23] satisfaction revealed a positive influence on behavioral intention. The largest coefficient was met among the three models indicating that satisfaction was the key antecedent when predict visitors' behavioral intention in wine tourism.

Beside, subjective norm has a positive influence on satisfaction and behavioral intention. This suggests that the effect of social approval and pressure will be more salient when the visitors‘ claims that a firm did not perform as promised, while the spreading of positive satisfaction depends mainly on the assessment of whether such behavior is appropriate. The findings were consisted with previous studies [27-29].

In particularity, involvement has a negative moderated effect on satisfaction - behavioral intention relation in wine tourism. These findings are similar to those by [18,34]. Given that visitors have high involvement with a wine tourism site, s/he is likely to immerse themselves in the experience of wine tourism site. S/he is likely to have better word of mouth, recommendation, and future visitation even s/he is not satisfied with the wine tour.

5. Conclusions

This study sets out to examine the influences of involvement, subjective norm, and satisfaction on behavioral intention with wine tourism sites, Yibin in China. Involvement and subjective norm directly result in a favorable satisfaction, which finally enhances to visitors' behavioral intention. Involvement moderates the relation between satisfaction and behavioral intention, while subjective norm does not.Consequently, the implications to smart tourism would be the importance of the equipments and service at Yibin region. For example, a visitor would have the higher probabilities to satisfy with wine tourist site when he/she can easily call the car or rent a bike to go around the heritage than those one who has difficulty to do that. Moreover, enhancing the cognition of the use for smart tourism tools would be important. Because the benefits of smart tourism tools can decline the constraints and fill the gap of the distance between visitor and destination through these dimensions: attractions, accessibility, amenities, available packages, activities, and ancillary services.

Findings indicate that increasing visitors' involvement in wine tourism sites directly improves visitors' satisfaction. In fact, the desire to be involved in the wine heritage experience is considered as one of the major motivations of wine heritage tourism. Thus, wine business should consider how to improve visitors' involvement.

In addition, this study not only empirically confirms previous assumptions that subjective norm is an influential factor altering visitors' satisfaction, but also develops the underlying rationale that it positively influences behavioral intention via satisfaction in wine tourism. This indicates that the subjective norms should be considered when predicting visitor satisfaction and behavioral intention in wine tourism context. Extending previous findings to a wine tourism context, this study provides additional evidence on the critical role of subjective norms.

Moreover, involvement moderates the relation between satisfaction and behavioral intention indicating that the higher the visitor level of involvement, the more detailed information on wine tourism they wanted and the more effort they made in selecting a product/service, resulting in positive satisfaction; and compared to those with high level of involvement, visitors with low level of involvement showed a close relationship between their satisfaction and behavioral intention. On the other hand, visitors with high level of involvement will present the positive words, recommend the sites, and promote the wine tour not only they satisfied with wine tour. The other reasons should be concerned by the wine business and future researchers.

From a company‘s perspective, the most important variable that can strengthen behavioral intention is visitors‘ satisfaction, but in a modern, competitive society it is very hard to enhance satisfaction through differentiation of service/product quality. The reason is that visitors select goods or services that can provide them with the most benefits among various alternatives. Thus, verifying the possible antecedents of both satisfaction and behavioral intention, and offering the benefits can contribute to the wine business management.

This study notes that wine visitor satisfaction and behavioral intention had organic causation. It also verified that whereas satisfied visitors presented active intention. Enhancing visitor satisfaction for performance improvement is crucial, but even satisfied visitors may make switching behaviors and companies should make more of an effort to maintain existed customers rather than satisfied customers. To this end, companies need to perform diverse relationship marketing strategies such as cementing customer membership and issuing coupons and discount tickets, thereby strengthening behavioral intention.

The limitations of this study provide evidence for further research. In the future, a representative sample might be desirable. This study focuses on examining the involvement from the perspective of the winetourist. Further research could compare supply and wine production perspectives to consumers' perceptions to generate an integrated understand in gand produce insightful contributions. Also, it might be interesting to conduct a qualitative study using a focus group of repeated visitors to understand their involvement and subjective norms to return more deeply. The findings from this study are on wine tourists from the Yibin wine regions. Thus, future studies should cover other wine regions to gain a broader spectrum of wine tourist behavior. Moreover, they should focus on the issues that why social reference is not significant in influencing visitor's intention. Further, more illustration of its impacts on behavioral intention in smart tourism context would be needed.

Appendix: measure scale

Involvement [5, 38]
INV1: I like being involved in making visits of wine sites
INV 2: I attach great importance to visiting wine sites
INV 3: Purchases decisions for wine tourism are very important to me
Subjective norms [27, 39]
SN1: People whose opinions I appreciate think that it is okay for me to visit this wine site.
SN2: Most people I value think that I should visit this wine site.
SN3: It is expected that I visit this wine site to see some wine products/ services. Satisfaction [38-40]
SA1: I am happy with the products/services I have visited on this wine site.
SA2: I am generally happy with having visited from this wine site.
SA3: In general, I have liked visiting this wine site.
Behavioral intention [10, 39, 43]
BI1: I say positive things about the experience at wine site.
BI2: I will recommend visiting wine site to anybody who seeks my advice.
BI3: I will revisit the wine site in the future.

References

[1] Li Y, Hu C, Huang C and Duan L2017 *T. Man.* **58** pp 293－300
[2] Koo C, Park J and Lee J N 2017 *E. Sma.Tou.* **54** pp 683－686
[3] Kim H, Lee JN and Han J 2010 *A.* **53** pp 151－156
[4] Gretzel U, Werthner H, Koo CandLamsfus C 2015 *H. Beh.* **50** pp 558－563
[5] Lu L, Chi CG and Liu Y 2015 *T. Man.* **50** pp 85－96
[6] Prayag G and Ryan C 2012 *J. of .Tra. Res.* **51**(3) pp 342－356
[7] ChingYuan H, ChiaJung C and PeiChing L 2010 *T. Man.* **31**(4) pp 513－526
[8] Brown G, Smith A and Assaker G 2016 *T.Man.* **55** pp 160－172
[9] Zaichkowski JL 1985 *J. Con. Res.* **12** pp 341－352
[10] Huang CY,Chou CJ and Lin PC 2010 *T Man.* **31**(4) pp 513－526
[11] Barber N, Taylor C and Strick S 2009 *I. J. Win Res.* **1** pp 59－72
[12] Kim YG, Suh BW and Eves A 2010 *J. Hos. Man.* **29** pp 216－226
[13] Campbell J M and DiPietro R B 2014 *J. Ret. Con. Ser.* **21**(5) pp 812－823
[14] Oliver R L 1999 *J. of.Mar.* **63** pp 33－34
[15] Baloglu S, Pekcan S, Chen L and Santos J 2004 *J. Qua. Ass. Hos. Tou.* **4**(3) pp 149－165
[16] Williams P and Soutar G N 2009 *A. Tou. Res.* **36**(3) pp 413－438
[17] Yoon Y and Uysal M 2005 *T. Man.* **26**(1) pp 45－56

[18] Mathis E F, KimHL, Uysal M, Sirgy J M and Prebensen N K 2016 *A.Tou.Res.* **57** pp 62－75
[19] Ali F, RyuK and Hussain K 2016 *J. Tra. Tou. Mar.* **33**(1) pp 85－100
[20] Kim H, Park J, Kim M and Ryu K 2013 *J. Hos. Man.* **33**(1) pp 397－405
[21] Jeon S andHyun S 2013 *C. Iss. Tou.* **16**(4) pp 343－368
[22] Prebensen NK, Woo E andUysal MS 2014 *I. Tour.* **17**(10) pp 910－928
[23] Han H and Hyun S S 2017 *I. Jou. Hos. Man.* **63** pp 82－92
[24] Oliver RL 1997 *Satisfaction: A Behavioral Perspective on the Consumer* (New York: McGraw-Hill)
[25] Tussyadiah I P 2016 *I. Jou. Hos. Man.* **55** pp 70－80
[26] Fishbein Mand Ajzen I 1975 *Beliefs, attitude, intention, and behavior: An introduction to theory and research* (Boston: Addison-Wesley.)
[27] Kim H B, Kim T T andShin S W 2009 *T. man.* **30**(2) pp 266－277
[28] SanMartín S, Prodanova J and Jiménez N 2015 *J. Ret. Con. Ser.* **23** pp 1－8
[29] Chen CC, Hsiao KL and Hsieh CH 2018 *I. Pre. Cor. Pro.* pp 1－8
[30] Cooke R andSheeran P 2004 *B. Jou. Soc. Psy.* **43** pp 159－186
[31] Olsen S O 2007 *P. Mar.* **24**(4) pp 315－341
[32] Caru A and Cova B 2007 *C. exp.* (New York. NY: Routledge)
[33] Baron S and Warnaby G 2011 *c. per.* (Boston, MA: Springer)
[34] Suh JC and Yi Y 2006 *J. Con. Psy.* **16**(2) pp 145－155
[35] Tuu HH and Olsen SO 2010 *A. Mar. Jou.* **18**(3) pp 151－158
[36] Hall C M and Macionis N 1998 *T. Rec. Rur. Are.* pp 197－224
[37] Yen TF 2018 *L. Stu. Res.* **16**(2) pp 1－15
[38] Jung H S and Yoon H H 2012 *I. Jou. Hos. Man.* **31**(3) pp 875－884
[39] Chiu P, Yen TF and Chu K 2015 *J. For. Agri.&Env.* **13**(3/4) pp 168－174
[40] Yen TF 2017 *C. MAT.* **123**(32) pp 1－5
[41] Anderson JC and Gerbing DW 1988 *P. Bul.* **103**(3) pp 411－423
[42] Lee J andBeeler C 2009 *E. Man.* **13**(1) pp 17－29

IOP Conf. Series: Materials Science and Engineering **658** (2019) 012011 doi:10.1088/1757-899X/658/1/012011

A development and application of maintainability evaluation system using immersive VR

Meihui Wang[1,2], *Yongxiang* Li[3], *Chuan* Lyu[1,2] and *Xi* Liu[4]

[1]School of Reliability and Systems Engineering, Beihang University, China
[2]State Key Laboratory of Virtual Reality Technology System, Beihang University, China
[3]Technology and Engineering Center for Space Utilization, Chinese Academy of Science, Beijing, China
[4]Xi'an Modern Control Technology Research Institute, Xi'an 710065, China

E-mail: lc@buaa.edu.cn

Abstract. Immersing into virtual reality (VR) is a perception of being physically present in a non-physical world. Using immersive VR in maintenance brings great conveniences. For example, using virtual human operating in hazardous environment before the real person, which can prevent being hurt. In this paper, a system of immersive virtual maintenance (IVM) is presented. It is divided into three parts. The first part is Input, which contains the maintenance task and maintenance design. The second part is Method, with which making the virtual human and the users act in a unified way. The key to build IVM system is the interface between the motion capture system and virtual simulation software. The third part is Output, which presents the maintainability evaluation results. This paper focuses on how to build IVM system, the difficulties and solutions also be detailed. At last, a simple application is given

1. Introduction

Maintainability is "the ease with which a software system or component can be modified to correct faults, improve performance or other attributes, or adapt to a changed environment" [1]. It is one of the main necessary factors not only constituting the system effectiveness but also affecting life cycle cost [2]. Traditional maintainability analysis and verification has some insufficiencies, for example, the defects can be detected only after the products have been put into use, which is difficult to modify the design [3].

Virtual maintenance provides a powerful tool to solve those problems in maintainability design, which improves the efficiency greatly and reduces the cost of design and future maintenance. Immersing into VR is a perception of being physically present in a non-physical world. The most superiority of the immersive VR system lies in that it allows the user to experience a virtual space built using computer graphics as with a real world [4].

The rapid development of the virtual reality technology makes a big progress in immersive virtual simulation [5]. Virtual humans have been used in many applications. Thorisson's [6] interactive guide, Gandalf, gives solar system tours. USC's Institute for Creative Technologies created virtual experiences to train soldiers in interpersonal leadership [7]. Just VR [8] allows a medical trainee to work with a virtual assistant to assess and treat a virtual victim. Balcisoy et al. [9] created a system where users play chess against a virtual human augmented to the real world. The Human Modelling and Simulation Group at the University of Pennsylvania uses virtual humans for task analysis and assembly validation [10].

Immersive virtual simulation plays a significant role in improving the product quality and saving the production cost; what's more, the virtual human can work in harsh conditions, which can avoid the accidents. Ramanarayan [11] studied virtual human's different types of walking. M. Susan Hallbeck [12] used the virtual simulation to formulate high efficiency and healthy postures. By integrating dynamic simulation and ergonomics evaluation, it enables the system designer to visualize and improve workplace design in the digital space [13]. Majid Hashemipour et al. [14] applied virtual technology in teaching.

However, the integration technologies of the motion capture system and the virtual simulation software does not get the satisfactory progress. The collected data from the motion capture is authenticity but lack of interaction and controllability, which makes the data reuse in the dynamic environment difficult [15]. The virtual maintenance simulation software has strong function in simulation analysis, but the work of virtual human motion modelling is tedious and the edit action is not real enough [16]. Therefore, integrating the two systems could control of the virtual human, get more effective data, and realize the true sense of immersion.

In the quest for a functioning, cost effective design, maintainability is often neglected. However, the cost of poor maintainability and reliability can be extremely high and the accumulated costs over the life of a plant often exceed the initial capital cost. In the engineering application, the artificial maintenance exposes lots of problems. For example, in the design stage, the maintenance persons don't know whether it is convenience to maintain when the machine is malfunction. To solve these problems, the immersive VR is introduced in the maintenance. With the help of the IVM, we can do the work about maintainability analysis (ME) and verification in the design stage, which can help find the insufficiencies before being produced. What's more, in IVM system the real maintenance experience can be got, and the maintenance activities can be simulated.

In this paper, a system of IVM is presented. The main function goal of IVM is as follows.

1) Import the virtual prototype and simulate the maintenance environment.
2) Simulate and verify the maintenance activities.
3) Analyze the maintenance visibility, accessibly, operating space and the maintenance time.
4) Output the maintainability evaluation.

The IVM system is divided into three parts. The first is Inputting Part, which contains the maintenance task and maintenance design. The second is Method Part, with which to make the virtual human and the user act in a unified way, it gives us the main ideas of the method, and the key to build IVM system is the interface of motion capture (MC) system and virtual simulation software (VSS). The third is Output Part, which gives us the maintainability evaluation. This paper focuses on how to build IVM system, the difficulties and solutions also be detailed. And at last, the application of IVM is given.

2. Methods and materials

2.1. The system of IVM

The system is designed according to the function goal. As shown in Figure 1, the overall frame structure for the logic level is divided into three parts: input part, IVM method part and output part. Input part contains maintenance task and standards of ME. IVM method part contains MC, construction of VR scene, and the interface of MC system and VSS. The output part contains the record of maintenance activities, ME and modify advice.

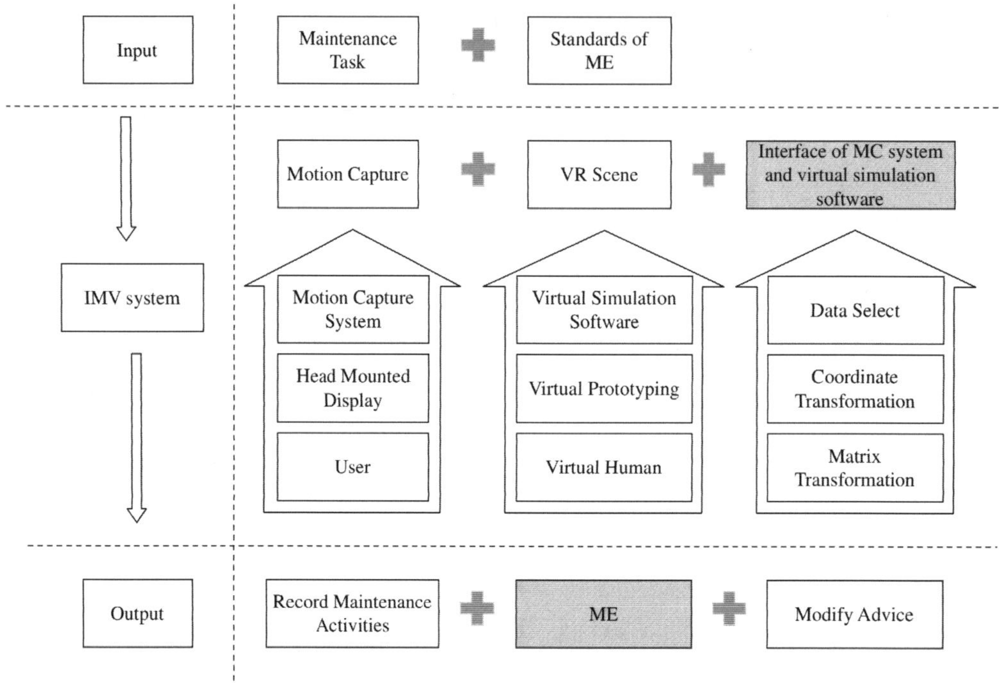

Figure 1. The structure of IVM system.

2.2. The basic definitions of IVM

In order to better understand this article, here we give some definitions:

Virtual Environment (VE): A three-dimensional model of a space displayed to a human user from an egocentric point of view using real-time 3D computer graphics.

Virtual Reality (VR): The experience of being within a VE.

Immersion: The feeling of "being there" that is experienced in some VEs. A VE user is immersed when he feels that the virtual world surrounds him and has to some degree replaced the physical world as the frame of reference.

Virtual Maintenance: The virtual maintenance is based on computer technology and virtual reality technology. In the VE, which contains product digital prototyping and virtual human model, the virtual human is driven to complete the repair process simulation. Virtual maintenance belongs to simulation, but it stresses the sense of reality.

Immersive Virtual Maintenance (IVM): The maintenance personnel wear the MC equipment, while the motion data is loaded into the computer and control the virtual human completing the repair process simulation.

Motion Capture (MC): Motion capture is the process of recording the movement of objects or people. It is used in military, entertainment, sports, medical applications, and for validation of computer vision [17] and robotics.

Head Mounted Display (HMD): a display device, worn on the head, which has a small display optic in front of each eye [18]. The HMD, Virtual Research VR 1280, is used in this paper.

2.3. The method of IVM

As shown in Figure 1, to build the IMV system, the maintenance task and the standards of ME should be firstly prepared. The key components of IVM system are MC, VR scene and Interface, as shown in Figure 2. The MC here is the user, dressed the MC system and head mounted display, to realize the feeling of immersive in the VR. The VR scene consists of virtual human, virtual prototyping, and VSS. The key to build IVM system is the interface of MC system and VSS, which help the virtual human and the user act in a unified way. The Output of the IMV is the Record, Modify Advice, and ME.

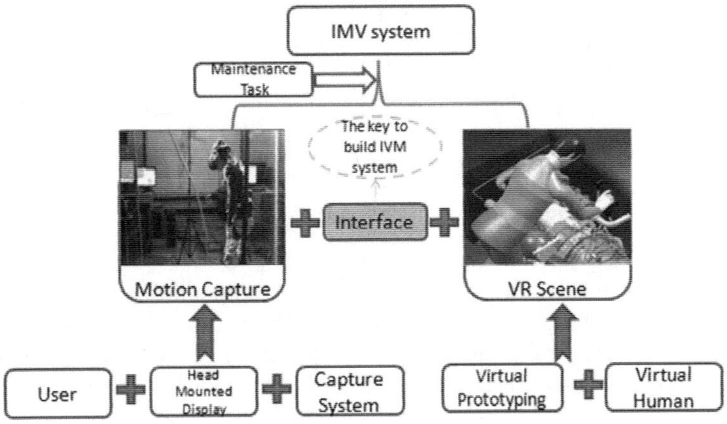

Figure 2. The key components of IVM system

2.1.1. *Motion capture*

2.1.1.1. The key components of MC. The MC here is composed of the user, the MC system and the head mounted display. The user dressed the MC system can translate the maintenance motion into data. The head mounted display is used to make the user immersing in VE.

We choose the ShapeWrap as the MC system. The ShapeWrap is a portable and durable motion captures system. Freedom from studios and cameras combined with its robust design make ShapeWrap ideal for use in the field, classroom or just about anywhere.

Because of its Wireless capabilities, the ShapeWrap can be operated away from the computer that receives the MC data. ShapeWrap conforms to fit almost any size person, and no special suit is required. Many researchers have used this MC system to capturing human motion and get credible action data [19-22].

2.1.1.2. The analysis of position and orientation of virtual human in ShapeWrap. In the GPO (Global Position and Orientation) data format from ShapeWrap, all limb data is stored with reference to the world coordinate system. All positional data is measured in mm and all orientation data is measured in roll, pitch and yaw orientation angles in degrees, which is shown in Figure 3. The X-axis extends outwards in the direction that the person is facing during the homing calibration, the Y-axis extends upwards, and the Z-axis extends to the person's right during homing. The order of rotation angles is yaw first about the +Y-axis, followed by pitch about the rotated +Z-axis, and then roll about the doubly rotated +X-axis. The default limb position (all rotation angles set to zero) is for a limb extending outwards along the X-axis. In other words, if a certain part of human manikin' posture is changed, we can obtain a series corresponding freedom degree.

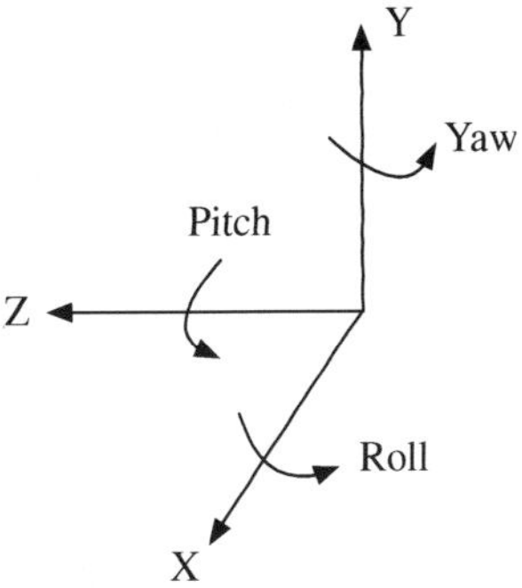

Figure 3. The global coordinate system in ShapeWrap

The Local coordinate systems are shown in Figure 2, the red line is +X-axis, the green line is +Y-axis, and the blue, and the blue line is +Z-axis. During the homing calibration, the local coordinate systems are not all the same with the global coordinate system. The coordinate of head, arm, hand and lumber is the same with the global, while the leg's coordinate system is around the Z-axis rotated 90 degrees clockwise.

2.1.2. VR scene

2.1.2.1. The key components of VR scene. VR scene is built in the VSS. The product CAD data is imported in VSS and shown as virtual prototyping. We use DELMIA, Digital Enterprise Lean Manufacturing Interactive Application, as the VSS. DELMIA is one of VSS from Dassault Company in France, has powerful analysis functions (Mohammad et al., 2010). The virtual human in the DELMIA is treated as a product, and its position and orientation is analysed as follows.

2.1.2.2. The analysis of position and orientation of virtual human in DELMIA. Like the ShapeWrap, the DELMIA also has its GPO, and all positional data are measured in mm and all orientation data is measured in roll, pitch and yaw, which are shown in Figure 4. But there are some differences between the two systems. Firstly, in the DELMIA, not all limb data is stored with reference to the world coordinate system, the arms' position and orientation is different from the global position and orientation, which is shown in Figure 4. Secondly, in ShapeWrap, the orientation angles data are saved in one array, while in DELMIA, the data are saved in two arrays, one for human attitude data and the other is for the orientation and position data.

In the GPO, the X axis extends outwards in the direction that the person is facing during the homing calibration, the Y axis extends to the person's left during homing, the Z axis extends to the upwards. The order of rotation angles is yaw first about the +Z-axis, followed by pitch about the rotated +Y-axis, and then roll about the doubly rotated +X-axis.

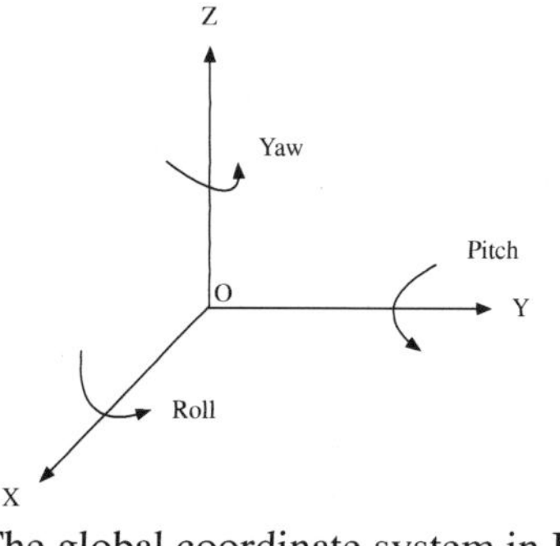

Figure 4. The global coordinate system in DELMIA.

2.1.3. Interface. The purpose of interface is to make the virtual human and the user act in a unified way, so that the maintenance activities can be evaluated in the VR scene. To obtain the interface, firstly, we should know the conversion matrix of MC system and VSS. Secondly, the conversion of coordinates between the MC system and VSS should be analysed.

2.1.3.1. The conversion matrix of MC system and VSS. The robotics conversion matrix is applied in the algorithm of software interface design, which is shown as Formula 1.

$$T = \begin{bmatrix} t_{11} & t_{12} & t_{13} & p_x \\ t_{21} & t_{22} & t_{23} & P_y \\ t_{31} & t_{32} & t_{33} & P_y \end{bmatrix} \quad (1)$$

The first three columns of the matrix represent the object physical coordinates position. In the global coordinate system, the Euler angles rotated along the +X-axis, +Y-axis and Z-axis are yaw, pitch and raw. The fourth column tells the coordinates position of the objects. The human body orientation defined by this array, and the above 3X3 in the array represents the orientation and the rest 1X3 means the human's position.

2.1.3.2. The analysis of the coordinate system. Comparing the coordinate system of the MC system and VSS, which is shown in Table 1, we can get that the different between them.

Table 1. Comparison of two kinds of software coordinate system.

Position	ShapeWrap	DELMIA
The global coordinate system (including neck, body, hips, leg, etc.)	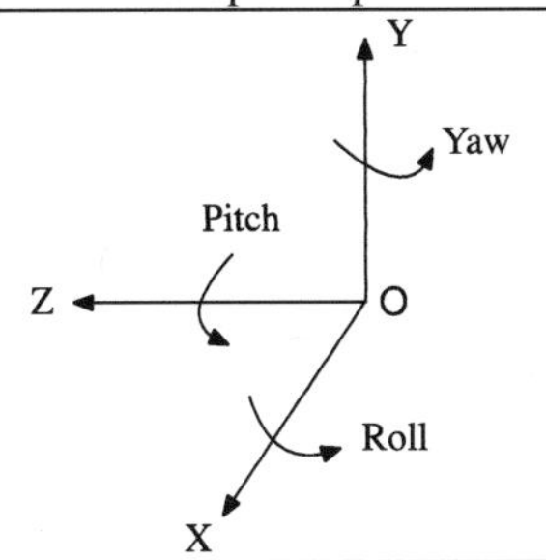	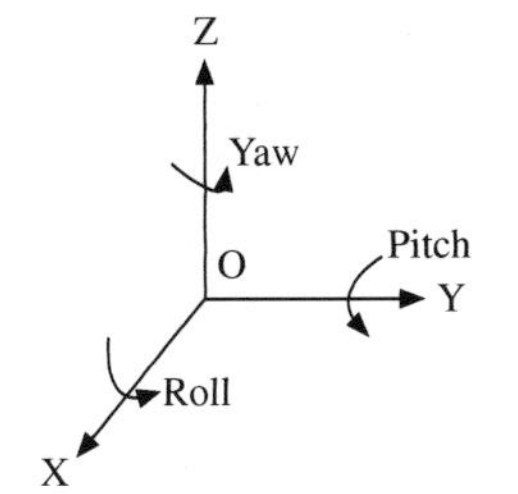
Shoulder (Right)	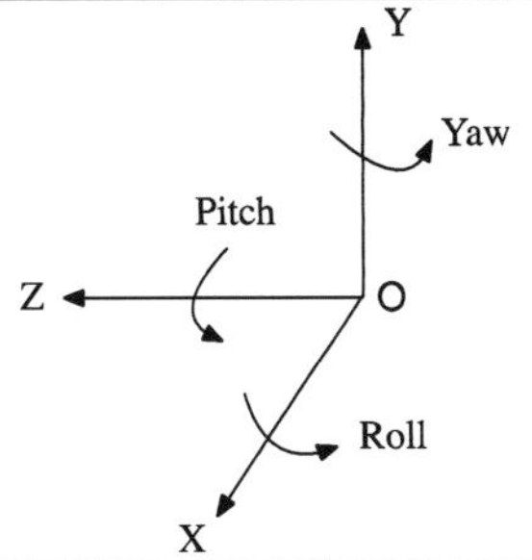	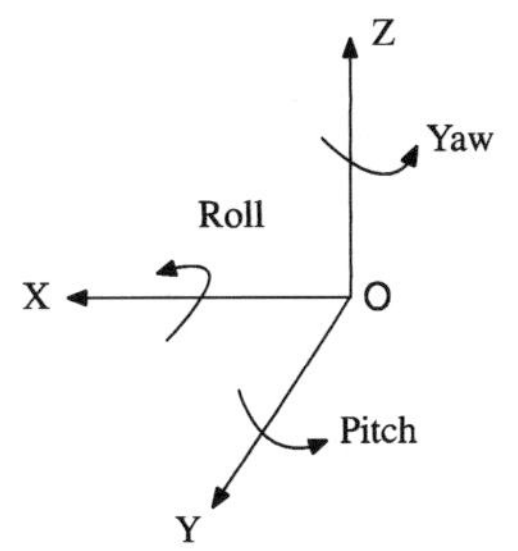
Arm (Right)	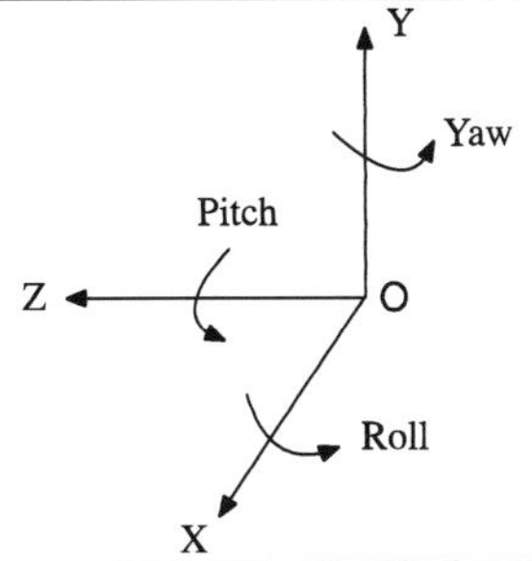	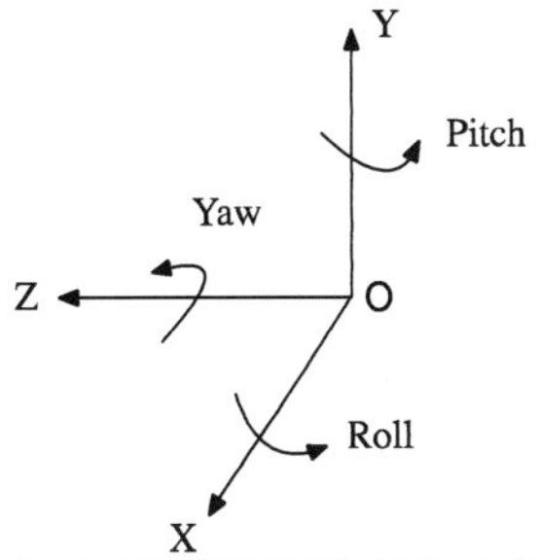
Ankle	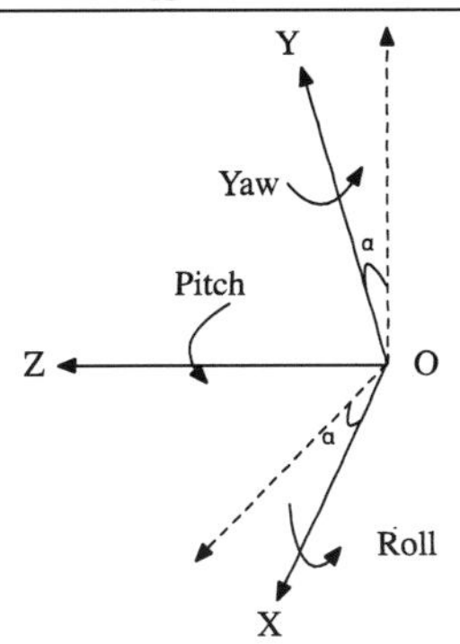	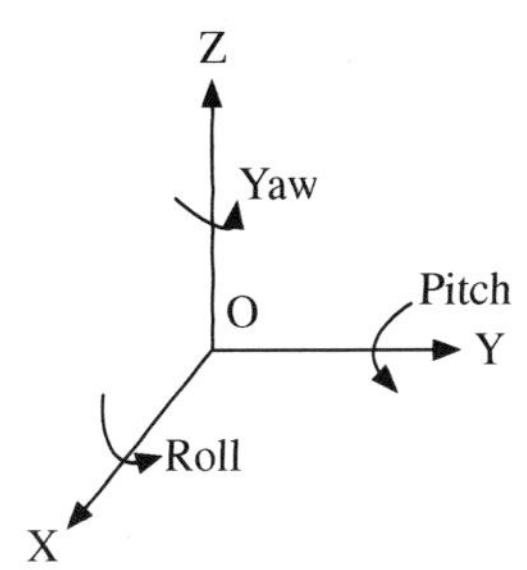

2.1.3.3. The interface algorithm description. The interface algorithm is mainly about the data conversion. The data conversion from MC system to VSS is shown in Figure 5.

Step 1 Get the GPO data from the MC software. ShapeWrap has two kinds of data output mode, one is the real-time output, and the other is file output. Here we choose the real-time data output. There are four kinds of data output format, which are GPO, C3D, BVH and Motion Builder, here we choose the GPO.

Step 2 Distinguish the GPO data into Particle data and Rigid Body data. The GPO data contains two kinds of data: Particle data and Rigid Body data, before data conversion, the Particle data and Rigid Body data should be distinguished.

Step 3 Data Conversion. The GPO data from the MC system cannot be directly used in the VSS, since the GPO coordinate and numbers are not same.

Convert the GPO data into the data that DELMIA can use, and store the data in the Array OPV and Array iAxis. The OPV is used to save the data of human joints curvature data and the iAxis is used to save the orientation and position data. The detailed calculation is in the Chapter 3.1.

Step 4 Assign the converted data to DELMIA. After assignment, the virtual human's motion in DELMIA is as same as the user dressed in ShapeWrap, which is shown in Figure 6.

Step 5. Repeat step 2 to step 4. Repeat step 2 to 4, until the MC system does not send the data to DELMIA.

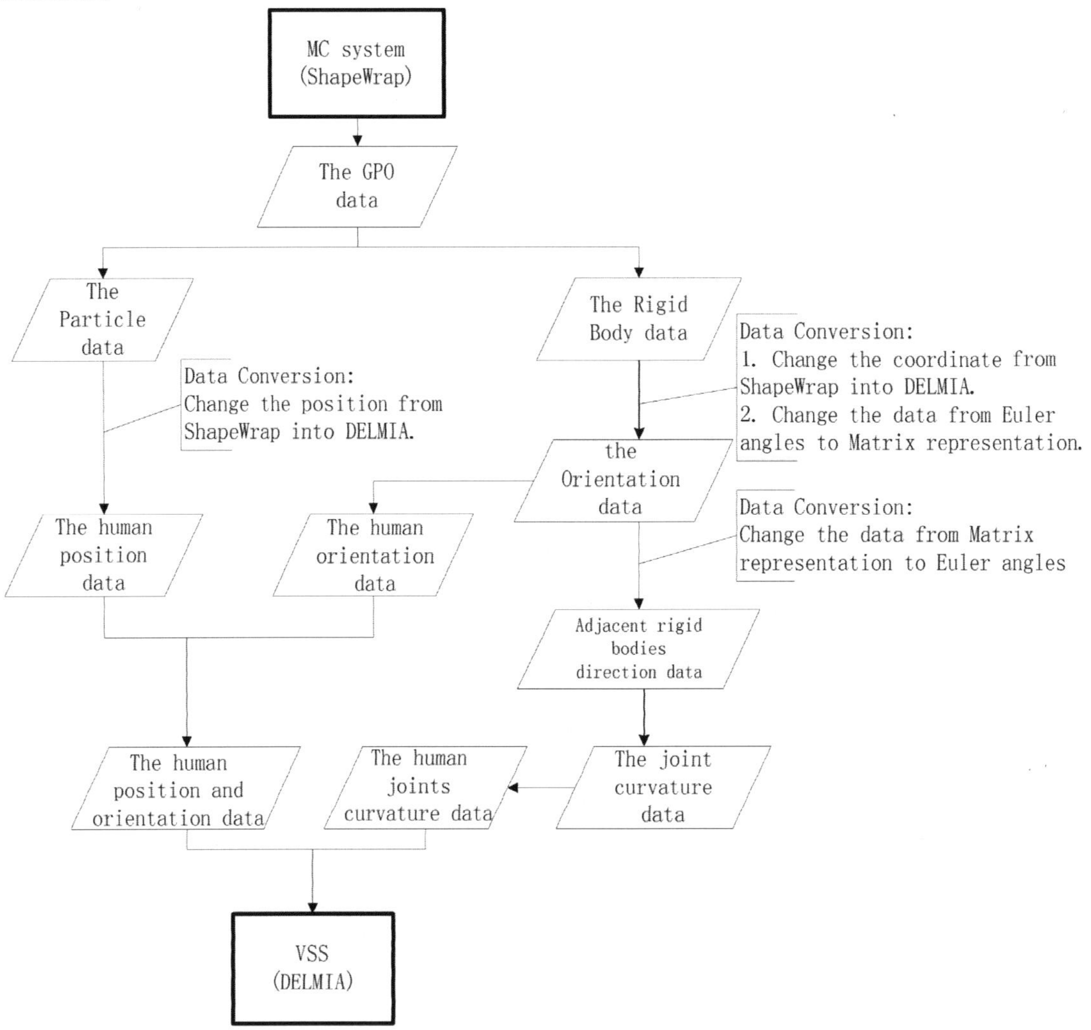

Figure 5. The data conversion from MC system to VSS.

Figure 6. The practical application of the simulation system.

2.4. Output

The output of IVM system contains video recording, ME, and modify advice. The video recording is used to teaching. The modify advice is according to conclusion of ME. The ME is divided into qualitative evaluation and quantitative evaluation. Using IVM system can more easily and friendly analyze the maintenance visibility, accessibility, and operating space, and also get more accurately maintenance time. The ME of IVM consists of qualitative evaluation and quantitative evaluation. The qualitative evaluation contains visibility, accessibility, operating space and the human comfort degree during maintenance. The quantitative evaluation is to evaluate the maintenance time.

1) The maintainability qualitative evaluation

The criterion of visibility is shown in Figure 7.1. The best vision means the user's vision is good during the maintenance. The max vision means the target can be seen, but may not be seen during the maintenance. Invisible vision means the maintenance target is not seen.

The criterion of accessibility is shown in Figure 7.2. In area 1, the maintenance target can be reached by easily, and the accessibility is good, while in area 2, the accessibility is bad.

Operating space means the space should be enough when do the maintenance task. The red circles shown in Figure 7.3 indicate there has collision between the hand and machine. The more the number of collisions, the less the operating space is.

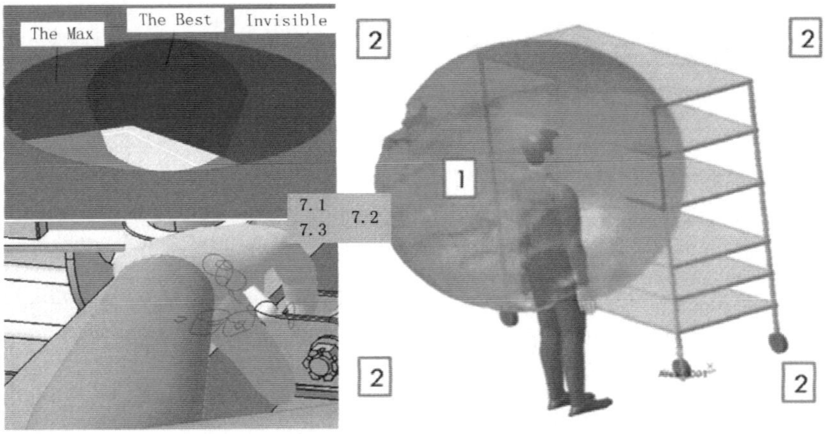

Figure 7. The criterion of visibility, accessibility and operating space in IVM.

The human comfort degree during maintenance is shown as Table 2, the upper-arm, forearm, hand, neck and trunk are analyzed. The score range is given, and each score is match is special color.

Table 2. The criterion of comfort degree in IVM.

	Score Range	Score and Color					
		1	2	3	4	5	6
upper-arm	1-6	Green	Green	Yellow	Yellow	Red	Red
Forearm	1-3	Green	Yellow	Red			
hand	1-4	Green	Yellow	Orange	Red		
Neck	1-6	Green	Green	Yellow	Yellow	Red	Red
Trunk	1-6	Green	Green	Yellow	Yellow	Red	Red

2) The maintainability quantitative evaluation

The maintenance time is evaluated here. In the IVM system, the length of time that the virtual human used is equal to the real maintenance time. At the beginning of the design, the length of the maintenance time is determined by the maintainability allocation. So, we can use the time obtained from IVM system to judge whether meet the requirements.

2.5. The difficulties and solutions

The conversion matrix is T, which is shown as in Formula 1, we use its first three columns as orientation T_0, and we get:

$$B = A.T_0 \tag{2}$$

Where A is the matrix of GPO coordinate in MC system, and B is the matrix of GPO coordinate in VSS. Both A and B is orthogonal matrix, and $|A| = |B| = 1$. Then the T_0 can be got as below:

1) Let the user, dressing the MC system, do the same action with the virtual human in the VSS, and remember the GPO data of A and B.

2) Through Formula 2, T_0 is obtained:

$$T_0 = B.A^{-1} = B.A^T \tag{3}$$

2.5.1. Calculate the virtual human's joints curvature in VSS. The joints curvature will be calculated with the data from MC system. Here, the right arm is taken as example, the orientation data of upper arm $iAxis_0$ and forearm $iAxis_1$ are given, use Array $iAxis_0$ and Array $iAxis_1$, we calculate the Array A, which contains the elbow angles data, and the angles data can be used in DELMIA.

Array A can be got from the follow equation.

$$\begin{aligned}
& A \cdot iAxis_0 = iAxis_1 \\
& \Leftrightarrow A = iAxis_1 \cdot iAxis_0^{-1} \\
& \Leftrightarrow A = iAxis_1 \cdot iAxis_0^{T} \\
& \Leftrightarrow A[0] = iAxis_1[0] \cdot iAxis_0[0] + iAxis_1[1] \cdot iAxis_0[3] + iAxis_1[2] \cdot iAxis_0[6] \\
& \quad A[1] = iAxis_1[0] \cdot iAxis_0[1] + iAxis_1[1] \cdot iAxis_0[4] + iAxis_1[2] \cdot iAxis_0[7] \\
& \quad A[2] = iAxis_1[0] \cdot iAxis_0[2] + iAxis_1[1] \cdot iAxis_0[5] + iAxis_1[2] \cdot iAxis_0[8] \\
& \quad A[3] = iAxis_1[3] \cdot iAxis_0[0] + iAxis_1[4] \cdot iAxis_0[3] + iAxis_1[5] \cdot iAxis_0[6] \\
& \quad A[4] = iAxis_1[3] \cdot iAxis_0[1] + iAxis_1[4] \cdot iAxis_0[4] + iAxis_1[5] \cdot iAxis_0[7] \\
& \quad A[5] = iAxis_1[3] \cdot iAxis_0[2] + iAxis_1[4] \cdot iAxis_0[5] + iAxis_1[5] \cdot iAxis_0[8] \\
& \quad A[6] = iAxis_1[6] \cdot iAxis_0[0] + iAxis_1[7] \cdot iAxis_0[3] + iAxis_1[8] \cdot iAxis_0[6] \\
& \quad A[7] = iAxis_1[6] \cdot iAxis_0[1] + iAxis_1[7] \cdot iAxis_0[4] + iAxis_1[8] \cdot iAxis_0[7] \\
& \quad A[8] = iAxis_1[6] \cdot iAxis_0[2] + iAxis_1[7] \cdot iAxis_0[5] + iAxis_1[8] \cdot iAxis_0[8]
\end{aligned} \tag{4}$$

A is a 3×3 array. The value of this array represents the forearm's orientation with respect to the upper arm. The coordinate system has 3 degrees, the 3 degrees' rotation sequence is assumed that α first, then β, and final γ.

$$\alpha = \arctan(-A[6] / A[8])$$
$$\beta = \arctan(-A[7])$$
$$\gamma = \arctan(A[1] / A[4])$$

Where α, β and γ represents the angle of yaw, the angle of pitch, and the angle of roll, respectively. And they are stored in OPV[b1], OPV[b2] and OPV[b3], which can be used as the elbow data in DELMIA.

2.5.2. Calculate the fingers' bending angle in VSS. Assume y, p and r respectively represent the roll angle, pitch angle and yaw angle, and then,

$$iAxis\_Hand = \begin{bmatrix} \cos y \cos p & \sin y \cos p & \sin p \\ -\cos y \sin p \sin r - \sin y \cos r & -\sin r \sin p \sin y + \cos r \cos y & \cos p \sin r \\ -\cos r \sin p \cos y + \sin r \sin y & -\cos r \sin p \cos y - \sin r \cos y & \cos p \cos r \end{bmatrix} \quad (5)$$

Assume GPO[a1], GPO[b1] and GPO[c1] represent F1 coordinate, and GPO[a2], GPO[b2] and GPO[c2] represent F2 coordinate, as shown in Figure 8.

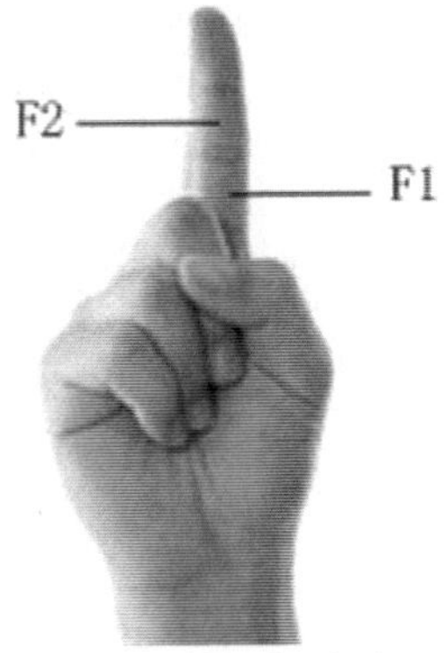

Figure 8. The finger that includes two joints.

$$dx = GPO[a2] - GPO[a1]$$
$$dy = GPO[c2] - GPO[c1]$$
$$dz = GPO[b2] - GPO[b1]$$

$$V[0] = dx \cdot iAxis\_Hand[0] + dy \cdot iAxis\_Hand[1] + dz \cdot iAxis\_Hand[2]$$
$$V[1] = dx \cdot iAxis\_Hand[3] + dy \cdot iAxis\_Hand[4] + dz \cdot iAxis\_Hand[5]$$
$$V[2] = dx \cdot iAxis\_Hand[6] + dy \cdot iAxis\_Hand[7] + dz \cdot iAxis\_Hand[8]$$

And at last, we get the data of finger joint between F1 and F2.

$$OPV[y] = \arcsin(V[1] / \sqrt{V[0]^2 + V[1]^2})$$
$$OPV[p] = -\arcsin(V[2] / \sqrt{V[0]^2 + V[2]^2})$$

OPV [y] and OPV [p] represent the yaw angle and pitch angle, and the finger's roll angle is not exit.

2.5.3. Holding tools in VSS. Maintenance tool is necessary during the actual maintenance operation, so is IVM. In order to get convincing ME results, the virtual human in VSS should hold the tools. To judge whether holding tools by virtual human in VSS is a big difficulty in IVM system. The algorithm is shown as Figure 9.

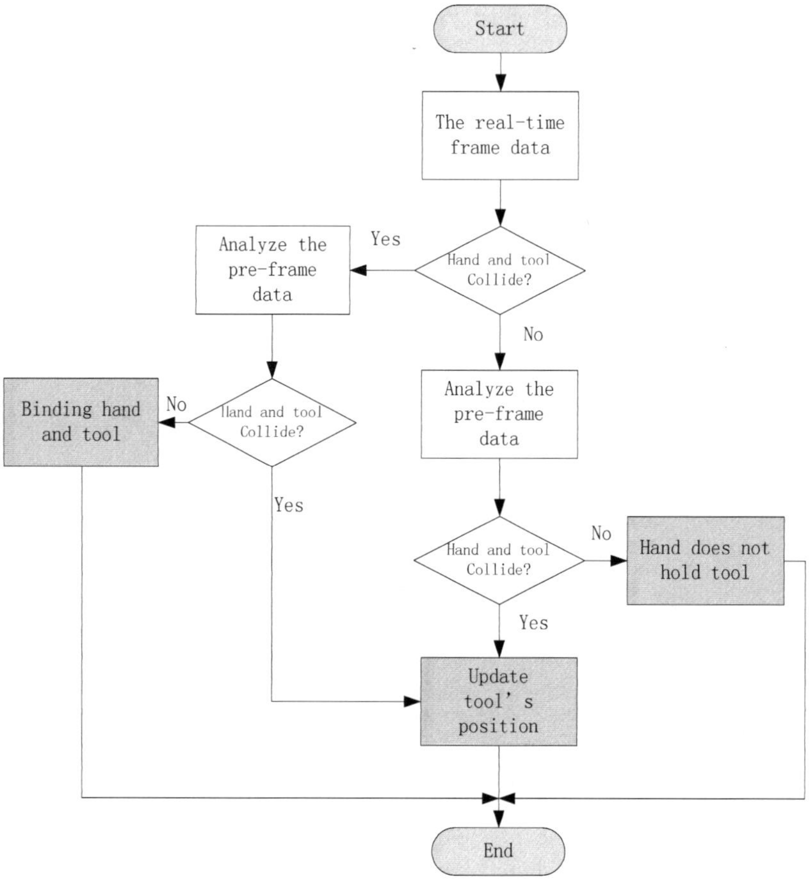

Figure 9. The process of judging whether holding tools.

Judging the real-time-frame data from the MC system, if collide, then use the pre-frame data to judge whether colliding, if yes, then update the tool's position with hand; if not, then bind hand with tool.

If using the real-time-frame data does not get colliding, then analyze the pre-frame data; if the hand collide with tool, then update the tool's position; if not, the hand would not touch the tool.

2.6. Case study

It is significant to introduce IVM system into the engineering maintenance. The user, dressing in the MC system and head mounted display, disassemble a component from a machine, which is shown in Figure 10.

The virtual human holds the wrench to screw. Through the Area 1(Figure 10.1), we know the accessibility and visibility is good, while the operating space is not big enough, the upper-arm is hit the machine.

In the Figure 10.2, the human comfort degree during maintenance is shown. The upper-arm's color is red and the forearm is Yellow. Compared with Table 2, the comfort degree is 3. In Figure 10.3, the user held this component and walk down the steps. The user's vision is good during the maintenance, and the visibility is the best.

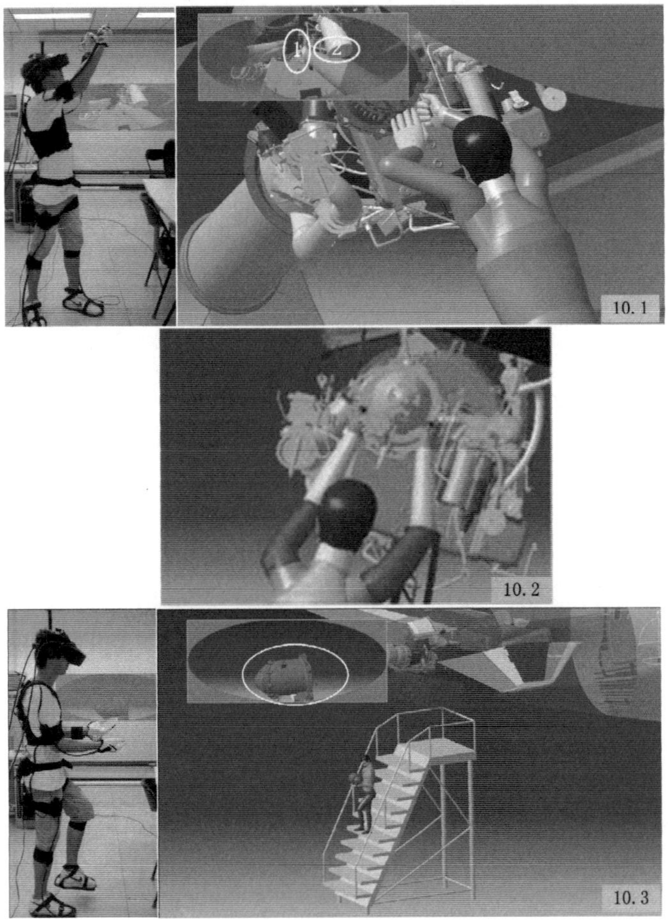

Figure 10. The qualitative evaluation in IVM.

The whole maintenance time in IVM is about 20 minutes, while the mean disassembly time allocated is 30 minutes. So the maintenance time to disassembly this component meets the requirements.

Through the IVM, we know the operating space to disassemble the component is not enough, and the comfort degree is needed to improve.

3. Conclusion

In this paper, a system of IVM is presented, and the key to build IVM system, the interface of MC system and VSS, is also detailed. IVM makes the maintenance design no longer totally dependent on the physical prototype or equipment. The possible maintenance design defects could be discovered in the early stage. IVM system can help the maintenance personnel operate the virtual product like in the real scene. Using IVM system can more easily and friendly analyze the maintenance visibility, accessibility, and operating space, and also get more accurately maintenance time, all of these make the ME more easily and accurately. In the future work, we will try our best to improve the accuracy of the interface, and study more effective ME method.

References

[1] IEEE Standard. Glossary of software engineering terminology 1990 *IEEE Std.* 610.12
[2] M Z Gan 1995 *Maintainability Design and Verification* (Beijing: National Defend Industry Press)

[3] Y H Yang, Z Z Li and L Zheng 2005 *J. Sys. Simul.* **17**(9) pp 2191 – 2198

[4] M Tomida and K Hoshino 2014 *Proc. of the 2014 IEEE/SICE Int. Symp. on System Integration* (Tokyo: Japan/IEEE) pp 163 – 168

[5] Michel Dureigne 2004 *Methods and Tools for Co-operative and Integrated Design* ed Serge Tichkiewitch and Daniel Brissaud (Netherlands: Springer) pp 3 – 14

[6] Thórisson K R 1997 *Proc. of the 1st Int. Con. on Autonomous agents* (Marina del Rey, California: USA) pp 536 – 537

[7] R W Hill Jr., J Gratch, S Marsella, J Rickel, W Swartout and D Traum 2003 *J. KI* **17**(4) pp 5 – 10

[8] Manganas M Tsiknakis, E Leisch, M Ponder, T Molet, B Herbelin, N Magnenat-Thalmann, D Thalmann, M Fato and A Schenone 2004 *J. Info. Tech. in Healthcare* **2**(6) pp 399 – 412

[9] S Balcisoy, R Torre, M Ponder, P Fua and D Thalmann 2000 *Proc. of Comp. Graphics Int.* (Geneva/Switzerland) pp 303 – 307

[10] N I Badler, C A Erignac and Y Liu 2002 *Commun. ACM* **45**(7) pp 56 – 63

[11] Ramanarayan Vasudevan, Aaron Ames and Ruzena Bajcsy 2013 *Nonlinear Anal: Hybri. Syst.* **7**(1) pp 101 – 115

[12] M Susan Hallbeck, Tim Bosch, Gu J W, Van Rhijn and Frank Krause 2010 *Hum. Factor Ergon. Man.* **20**(4) pp 300 – 315

[13] Shao-Wen Chang, Mao-Jiun and J Wang 2007 *Hum. Factor Ergon. Man.* **17**(5) pp 445 – 455

[14] Majid Hashemipour, Hamed Farahani Manesh and Mert Bal 2011 *Comput. Appl. Eng. EDUC* **19**(2) pp 305 – 314

[15] Antoine Rennuit and Alain Micaelli 2005 *Proc. of Virtual Concept* (Biarritz/France: University Karlsruhe) pp 1 – 9

[16] Ming Yang and Mingde Yin 2009 *J. Anhui Agr. Sci.* **7** pp 12 – 15

[17] David Noonan, Peter Mountney, Daniel Elson, Ara Darzi and Guang-Zhong Yang 2009 *IEEE Int. Conf. on Robotics and Automation* (Kobe/Japan) **6** pp 3203 – 3838

[18] Jane M Lawrence, Kamyar Abhari, Steven L Prime, Benjamin P Meek, Loni Desanghere, Lee A Baugh and Jonathan J Marotta 2011 *Behav Res Methods* **43**(2) pp 399 – 408

[19] Dong Zhou, Zhiqi Guo, Meihui Wang and Chuan Lv 2011 *Appl. Mech. and Mater* 2011 **44-47** pp 532 – 536

[20] Baocheng Liu, Yongxiang Li and Dong Zhou *2013 Int. Conf. on Quality, Reliability, Risk, Maintenance, and Safety Engineering* (Chengdu/China) **7** pp 1986 – 1988

[21] Mohammad, Vipul Babriya and Tarek Sobh 2010 *Tech. Develop. in Edu. and Autom* pp 549 – 558

[22] Jurgen Bernard, N Wilhelm, B Kruger, T May, T Schreck and J Kohlhammer 2013 *IEEE T. Vis. Comput. Gr.* **19**(12) pp 2257 – 2566

IOP Conf. Series: Materials Science and Engineering **658** (2019) 012012 doi:10.1088/1757-899X/658/1/012012

Economic manufacturing quantity model with machine breakdown and deteriorating production process

Chung Ho Chen

Department of Industrial Management and Information, Southern Taiwan University of Science and Technology, Taiwan

E-mail: chench@stust.edu.tw

Abstract. In this study, the author proposes a jointly setting model for determining the process mean and economic manufacturing quantity (EMQ) under the machine breakdown and deteriorating production process. The system addresses the corrective maintenance, preventive maintenance, and allowable shortage. The quality loss of conforming product is considered and Taguchi's asymmetric quadratic quality loss function is used for evaluating the product quality. The optimal process mean and economic manufacturing quantity are jointly determined by minimizing the total expected cost of product per unit time including the set-up cost, holding cost, corrective maintenance cost, preventive maintenance cost, shortage cost. A solution procedure is devised to obtain the optimal solution and the sensitivity analysis of key parameters is conducted to investigate the effect on the optimal solution.

1. Introduction

Recently, the supply chain management with quality and reliability design is an important topic for the operational management on the manufacturing/service industry. For the two-level supply chain system, the trade-off problem between the manufacturer and customer should be available for further study. Process mean (process quality level) and production quantity settings are two different problems in the modern quality and inventory management. For the economic manufacturing quantity (EMQ) decision, we need to consider the minimum inventory cost including the set-up cost and holding cost. For the process mean decision, we need to consider the trade-off cost between the conforming and non-conforming product. We can solve the above individual model when the management problem occurs. If we would like to obtain the efficient enterprise resource usage, then we can consider to simultaneously determine the product and process parameters. Hence, we should prefer to adopt the integrated model for solving the above two management problems.

Traditional EMQ model assumes that the perfect product for the manufacturing process. Hence, the defective product for EMQ model has been neglected. Previous researchers, e.g., Porteus [1] and Rosenblatt and Lee [2-3], firstly proposed the imperfect quality of EMQ model. Recently, EMQ model with imperfect quality, inventory control, reliability, and supply chain is the important topic of business, management, and engineering.

There are some works that assured a post-sale warranty cost where imperfection is prevalent in the production system. The works of Djamaludin et al. [4], Yeh and Lo [5], Yeh, et al. [6], Wang and Sheu [7-8], Wang [9-10], and Yeh and Chen [11] can be quoted along this line. Recently, Mujahid and Rahim [12], Rahim and Fareeduddin [13], and Akram et al. [14] have proposed the integrated determination of production run length and product quality control problem. Sarkar et al. [15], Sana et al. [16], and Sana [17] have addressed the production lot size, sales price, and process maintenance control problem. Lin and Gong [18] considered an integrated manufacturer-buyer supply chain model for vendor's production system with random breakdown. When a bereakdon occurs, it needs to perform the correct maintenance to restore the production system.Economic selection of optimal process mean is a popular topic for modern statistical process control because it has a significant effect on the expected profit/cost per unit product. The study of process mean problem started from Springer [19]. Taguchi [20] redefined the product quality as the loss of society when the product is shipped and sold to the customer. To decrease the society's loss, he proposed the quadratic quality loss function for evaluating the product quality. Taguchi's [20] quality loss function can combine with the on-line quality control (production process control) and off-line quality control (product quality design) methods and promote the product quality and performance. His quality loss function has been successfully applied in industry for promoting the product quality and reliability by reducing the product cost. Jeang [21] proposed an integrated model with optimal production quantity, process mean, and process tolerance under the deterioration and breakdown production process. Recently, Chen and Tsai [22] and Chen et al. [23-24] have adopted Taguchi's [20] quality loss function for determining process mean and production quantity. Their models addressed that the economic model of product will underestimate the expected total relevant cost because the quality loss of conforming product is neglected. Many works about optimum process mean setting problem have been integrated with inspection plan, production and inventory control, maintenance, and supply chain management.

In 2010, Sarkar et al. presented a production system with machine failure and production process deterioration. Their model considers shortage and with/without safety stock for determining the optimal machine reliability parameter and production lot size. However, their model did not consider the quality loss of conforming products. In this study, the author proposes a modified Sarkar et al.'s [15] model with product quality loss for minimizing the total relevant cost of product per unit time. Assume that the quality characteristic of the product is normally distributed. Taguchi's [20] asymmetric quadratic quality loss function will be used for measuring the product quality. The total relevant cost of product per unit time for the modified model includes the set-up cost, corrective cost, preventive cost, holding cost, shortage cost, and expected quality loss of the conforming products and expected cost of the non-conforming products. To our best knowledge, the modified Sarkar et al.'s [15] model with quality loss of the conforming products is a new method for extending the application of Sarkar et al.'s [15] model. The main difference between our modified model and the original Sarkar et al.'s [15] one is that the former addresses the optimal product quality output for the production process and hence should include the trade-off between the quality loss and production cost. The important contribution of the this study is that the determination of optimal production and product parameters is based on the minimization expected total loss of society including the producer's manufacturing cost and customer's use cost for product. The management implication of this work is that the manufacturer provides a high quality and reliability product/service will promote the customer's satisfaction and reduces the total expect cost of the supply chain system. A solution procedure is devised to find the optimal solution for the modified Sarkar et al.'s [15] model and the sensitivity analysis of key parameters is conducted to investigate the effect on the optimal solution.

2. Original Sarkar et al.'s model for determining the reliability parameter and production quantity

Sarkar et al. [15] addressed the unreliable production system with process deterioration, machine breakdown and repair (corrective and preventive maintenance). During long-run process, a certain percentage of defective items are produced which is reworked at a cost immediately. The expected cost function including the corrective cost, preventive cost, set-up cost, holding cost, shortage cost, and rework cost is formulated while shortage for breakdown and with/without safety stock. The product lot size and reliability parameter of the machinery system are considered to be two decision variables. The aim of their model is to generalize economic manufacturing model with unreliable production system by taking into account jointly the process deterioration, machine breakdown, and maintenance policy.

3. Modified Sarkar et al.'s model for determining the process mean and production quantity

3.1. Notations

D	the demand rate in units per unit time
p	the production rate in units per unit time, $p > D$
C_h	the holding cost per unit per unit time
C_0	the setup cost for each production run
C_1	the corrective repair cost per unit time
C_2	the preventive repair cost per unit time
C_j	the scrap cost for a defective item
C_r	the rework cost for a defective item
C_m	the material cost per unit product
C_s	the shortage cost per unit product
Q	the economic manufacturing quantity
$C(\mu_y, Q)$	the expected total relevant cost of product per unit time
$S(\mu_y, Q)$	the expected total relevant cost of product
$L(Q)$	the expected cycle length
ρ	the elapsed time until production process shifts, it is an exponential distribution with parameter k
$f(\rho)$	the probability density function of ρ
T	the elapsed time until the machine fails, it is an exponential distribution with parameter ϕ
$f(t)$	the probability density function of T
X_1	the elapsed time until the corrective repair completes, it is an exponential distribution with parameter λ_1
$f(x_1)$	the probability density function of X_1
X_2	the elapsed time until the preventive repair completes, it is an exponential distribution with parameter λ_2
$f(x_2)$	the probability density function of X_2
Y	the normal quality characteristic of product
$f(y)$	the probability density function of Y
μ_y	the process mean (process quality level)
σ_y	the process standard deviation
LSL	the lower specification limit of product
USL	the upper specification limit of product
$\Phi(\cdot)$	the cumulative distribution function for the standard normal random variable
$z(\cdot)$	the probability density function for the standard normal random variable
α	the probability of defective product
$E(N)$	the expected number of products in a production cycle
$E_1(N)$	the expected number of rework items in a production cycle

$E_2(N)$	the expected number of scrap items in a production cycle
$E_3(N)$	the expected number of conforming items in a production cycle
y_0	the target value of product
$E[Loss(Y)]$	the expected quality loss per unit product
k_1	the quality loss coefficient when $LSL < Y < y_0$
k_2	the quality loss coefficient when $y_0 < Y < USL$
Δ_1	the deviation from the left-hand side of the target value of the product characteristic
Δ_2	the deviation from the right-hand side of the target value of the product characteristic

3.2. Assumptions

There are some assumptions in the modified model:

1. The elapsed time of production process shifting from an in-control state to an out-of-state, ρ, is an exponential distribution with parameter k and probability density function $f(\rho)$.
2. The elapsed time until the machine fails, T, is an exponential distribution with parameter ϕ and probability density function $f(t)$.
3. The elapsed time until the corrective repair completes, X_1, is an exponential distribution with parameter λ_1 and probability density function $f(x_1)$.
4. The elapsed time until the preventive repair comples, X_2, is an exponential distribution with parameter λ_2 and probability density function $f(x_2)$.
5. The quality characteristic of product, Y, is normally distributed with known standard deviation σ_y and unknown process mean μ_y.
6. The product is reworked when the quality characteristic is above the *USL*.
7. The product is scrapped when the quality characteristic is below the *LSL*.
8. Taguchi's asymmetric quadratic quality loss function is adopted for measuring the product quality.
9. Single item is produced in a single machine. During the production run time, the random breakdowns may occur. Only one breakdown in one cycle is considered.
10. If the machine breakdown occurs, then the corrective maintenance starts immediately; otherwise, the preventive maintenance is started as on time.
11. Shortage is considered for the corrective/ preventive repair periods.
12. The production rate and demand rate are considered as fixed.
13. The unreliable manufacturing system results in a machine breakdown in which the production is interrupted.
14. The unreliable production system takes into account jointly the process deterioration, machine breakdown, and repair (corrective and preventive maintenance).
15. There is no safety stock.

3.3. Mathematical model

Similar to Sarkar et al. (2010), the corrective/preventive repair time considers the following four periods: (1) the corrective repair time occurs between 0 and $\frac{p-D}{D}\cdot t$; (2) the corrective repair time occurs beyond $\frac{p-D}{D}\cdot t$; (3) the corrective repair time occurs between 0 and $\frac{p-D}{D}\cdot\frac{Q}{p}$; (4) the preventive repair time occurs beyond $\frac{p-D}{D}\cdot\frac{Q}{p}$. In Sarkar et al.'s [15] model, the total relevant cost of product for one production cycle includes the set-up cost, material cost, corrective repair cost, preventive repair cost, holding cost, shortage cost, and expected non-conforming cost of conforming. However, their model does not consider the expected quality loss of conforming products. Hence, Sarkar et al.'s [15] model will underestimate the expected total relevant cost of products. Our modified model should address the conforming cost of products. Similar to Sarkar et al. [15], the total relevant cost of product per unit time is

$$C(\mu_y, Q) = \frac{S(\mu_y, Q)}{L(Q)} \tag{1}$$

Where

$$\begin{aligned}
&S(\mu_y, Q) \\
&= C_0 + C_m Q + \frac{C_1}{\lambda_1}[1-\exp(-\phi\frac{Q}{P})] + \frac{C_2}{\lambda_2}\exp(-\phi\frac{Q}{P}) + C_h\frac{p(p-D)}{2D}\{-(\frac{Q}{p})^2\exp(-\phi\frac{Q}{p}) \\
&+\frac{2}{\phi}[-\frac{Q}{p}\exp(-\phi\frac{Q}{P}) + \frac{1}{\phi}(1-\exp(-\phi\frac{Q}{p}))] \\
&-\frac{\phi}{\phi+\lambda_1\frac{p-D}{D}}\{-(\frac{Q}{p})^2\exp[(-(\phi+\lambda_1\frac{p-D}{D})Q/p] \\
&+\frac{2}{\phi+\lambda_1\frac{p-D}{D}}\{-\frac{Q}{p}\exp[(-(\phi+\lambda_1\frac{p-D}{D})Q/p] \\
&+\frac{1}{\phi+\lambda_1\frac{p-D}{D}}[1-\frac{Q}{p}\exp[(-(\phi+\lambda_1\frac{p-D}{D})Q/p]]\}\} \\
&+C_h\frac{p(p-D)}{2D}(\frac{Q}{p})^2[1-\exp(-\lambda_2\frac{p-D}{D}\frac{Q}{p})]\exp(-\phi\frac{Q}{p}) \\
&+C_s D\frac{\phi}{\phi+\lambda_1\frac{p-D}{D}}[1-\exp[(-(\phi+\lambda_1\frac{p-D}{D})Q/p]] \\
&+C_s D\frac{1}{\lambda_2}\exp[(-(\phi+\lambda_2\frac{p-D}{D})Q/p]] + C_r p[1-\Phi(\frac{USL-\mu_y}{\sigma_y})]\{\frac{1}{\phi}[1-\exp(-\phi\frac{Q}{p})] \\
&+(\frac{\phi}{k(k+\phi)}-\frac{1}{k})[1-\exp(-(k+\phi)\frac{Q}{p})]\} \\
&+C_j p[\Phi(\frac{LSL-\mu_y}{\sigma_y})]\{\frac{1}{\phi}[1-\exp(-\phi\frac{Q}{p})] + (\frac{\phi}{k(k+\phi)}-\frac{1}{k})[1-\exp(-(k+\phi)\frac{Q}{p})]\} \\
&+E[Loss(Y)]p[\Phi(\frac{USL-\mu_y}{\sigma_y}) - \Phi(\frac{LSL-\mu_y}{\sigma_y})] \\
&+\{\frac{1}{\phi}[1-\exp(-\phi\frac{Q}{p})] + (\frac{\phi}{k(k+\phi)}-\frac{1}{k})[1-\exp(-(k+\phi)\frac{Q}{p})]\}
\end{aligned} \tag{2}$$

$$
\begin{aligned}
&E[Loss(Y)] \\
&= \int_{LSL}^{y_0} k_1(y-y_0)^2 f(y)dy + \int_{y_0}^{USL} k_2(y-y_0)^2 f(y)dy \\
&= k_1\{[(\mu_y - y_0)^2 + \sigma_y^2] \cdot \left[\Phi\left(\frac{y_0-\mu_y}{\sigma_y}\right) - \Phi\left(\frac{LSL-\mu_y}{\sigma_y}\right)\right] \\
&-\sigma_y[(\mu_y - y_0)z(\frac{y_0-\mu_y}{\sigma_y}) - (\mu_y - 2y_0 + LSL)z\left(\frac{LSL-\mu_y}{\sigma_y}\right)]\} \\
&+k_2\{[(\mu_y - y_0)^2 + \sigma_y^2] \cdot \left[\Phi\left(\frac{USL-\mu_y}{\sigma_y}\right) - \Phi\left(\frac{y_0-\mu_y}{\sigma_y}\right)\right] \\
&-\sigma_y[-(\mu_y - y_0)z(\frac{y_0-\mu_y}{\sigma_y}) + (\mu_y - 2y_0 + USL)z\left(\frac{USL-\mu_y}{\sigma_y}\right)]\}
\end{aligned}
\tag{3}
$$

Eq. (1) needs to find the optimal process mean μ_y^* and production quantity Q^* which minimizes the total relevant cost of product per unit time.

$$
\begin{aligned}
&L(Q) \\
&= \int_0^{Q/p} [\int_0^{\frac{p-D}{D}t} \frac{p}{D} t f(x_1)dx_1] f(t)dt + \int_0^{Q/p} [\int_{\frac{p-D}{D}t}^{\infty} (t+x_1) f(x_1)dx_1] f(t)dt \\
&+ \int_{Q/p}^{\infty} [\int_0^{\frac{p-D}{D}\frac{Q}{p}} \frac{Q}{D} f(x_2)dx_2] f(t)dt + \int_{Q/p}^{\infty} [\int_{\frac{p-D}{D}\frac{Q}{p}}^{\infty} (\frac{Q}{p} + x_2) f(x_2)dx_2] f(t)dt \\
&= \frac{p}{D}\frac{1}{\phi}[1-\exp(-\phi\frac{Q}{p})] + \frac{\phi[1-\exp(-\frac{Q}{p}(\phi+\lambda_1\frac{p-D}{D}))]}{\lambda_1(\phi+\lambda_1\frac{p-D}{D})} + \frac{\exp(-\frac{Q}{p}(\phi+\lambda_2\frac{p-D}{D}))}{\lambda_2}
\end{aligned}
\tag{4}
$$

3.4. Solution procedure

The mathematical optimization model for determining the decision variables (μ_y, Q) is to minimize $C(\mu_y, Q) = \frac{S(\mu_y, Q)}{L(Q)}$. One cannot prove that Eq. (1) is convex with respect to μ_y because of the cumulative distribution function of the standard normal random variable $\Phi(\cdot)$ for a given Q. Hence, we haven't the closed-form solution of μ_y for a given Q. The optimal solution of μ_y^* for a given Q is found numerically. A one-dimensional direct search method can be used to find the optimal process mean. Let $LSL < \mu_y < USL$. We can adopt the direct search method for solving μ_y^* with minimum $C(\mu_y, Q)$ under the given Q. Substituting the combination of (μ_y^*, Q) into Eq. (1), we have the responding total relevant cost of product per unit time.

The detail solution procedure for the above Eq.(1) is as follows:

Step 1. Give the maximum Q, Q_m.

Step 2. Let Q =1.

Step 3. Let $LSL < \mu_y < USL$. We adopt the direct search method for obtaining the optimal process mean μ_y^*. Substituting the combination of (μ_y^* , Q) into Eq. (1), we obtain the responding $C(\mu_y^*, Q)$.

Step 4. Let Q = Q+1. Repeat Step 3 until $Q = Q_m$. The combination of (μ_y^*, Q^*) with the minimum total relevant cost of product per unit time $C(\mu_y^*, Q^*)$ is the optimal solution.

4. Numerical example and sensitivity analysis

Some parameters are given as follows: $C_0 = 300$, $C_p = 10$, $C_1 = 0.5$, $C_2 = 5$, $C_h = 0.5$, $C_s = 1$, $C_j = 2$, $C_r = 3$, p = 240, D = 100, $\lambda_1 = 2$, $\lambda_2 = 10$, $\phi = 0.5$, k = 2, $\sigma_y = 0.5$, LSL = 6, USL = 10, $y_0 = 8$, $k_1 = 2$, and $k_2 = 4$. By solving the above Eq. (1), we have the optimal manufacturing quantity $Q^* = 133$ and process mean $\mu_y^* = 7.93$ with the total relevant cost of product per unit time $C(\mu_y, Q) = 1379.43$.

Table 1 lists the effect of some parameters on the optimal solution. From Table 1, one has the following conclusions: (1) the process standard deviation, two quality loss coefficients, upper specification limit, and target value of product have a major effect on the process mean; (2) the material cost, production rate per unit time, parameter of the elapsed time until the corrective repair completes, parameter of the elapsed time until the machine fails, process standard deviation, and upper specification limit have a major effect on the manufacturing quantity; (3) the material cost and demand rate per unit time have a major effect on the total relevant cost of product per unit time.

Table 1. The sensitivity analysis of parameters for numerical example.

C_1	μ_y	Q	$C(\mu_y, Q)$
0.4	7.93	133	1379.42
0.6	7.93	133	1379.44
C_2	μ_y	Q	$C(\mu_y, Q)$
4	7.93	133	1379.36
6	7.93	133	1379.49
C_h	μ_y	Q	$C(\mu_y, Q)$
0.4	7.93	134	1376.00
0.6	7.93	132	1382.82
C_s	μ_y	Q	$C(\mu_y, Q)$
0.8	7.93	133	1378.39
1.2	7.93	133	1380.46
C_j	μ_y	Q	$C(\mu_y, Q)$
1.6	7.93	133	1379.43
2.4	7.93	133	1379.43
C_r	μ_y	Q	$C(\mu_y, Q)$
2.4	7.93	133	1379.43
3.6	7.93	133	1379.43
C_p	μ_y	Q	$C(\mu_y, Q)$
8	7.93	147	1160.58
12	7.93	123	1595.18
p	μ_y	Q	$C(\mu_y, Q)$
192	7.93	120	1395.63
288	7.93	145	1362.89
D	μ_y	Q	$C(\mu_y, Q)$
80	7.93	133	1129.16
120	7.93	133	1616.65
λ_1	μ_y	Q	$C(\mu_y, Q)$
1.6	7.93	131	1353.23

2.4	7.93	135	1396.37
λ_2	μ_y	Q	$C(\mu_y,Q)$
8	7.93	133	1379.31
12	7.93	133	1379.41
ϕ	μ_y	Q	$C(\mu_y,Q)$
0.4	7.93	148	1356.37
0.6	7.93	122	1398.74
k	μ_y	Q	$C(\mu_y,Q)$
1.6	7.93	134	1375.45
2.4	7.93	133	1382.92
σ_y	μ_y	Q	$C(\mu_y,Q)$
0.4	7.94	136	1369.81
0.6	7.91	131	1390.92
k_1	μ_y	Q	$C(\mu_y,Q)$
1.6	7.91	134	1377.33
2.4	7.95	133	1381.43
k_2	μ_y	Q	$C(\mu_y,Q)$
3.2	7.95	134	1376.10
4.8	7.91	133	1382.66
LSL	μ_y	Q	$C(\mu_y,Q)$
4.8	7.93	133	1379.43
7.2	7.93	133	1379.49
USL	μ_y	Q	$C(\mu_y,Q)$
8	7.61	130	1395.61
12	7.93	133	1379.43
y_0	μ_y	Q	$C(\mu_y,Q)$
6.4	6.42	132	1384.79
9.6	9.40	132	1385.33

5. Conclusions

In this study, the author has proposes a modified Sarkar et al.'s [15] model for determining the process mean and manufacturing quantity under the machine breakdown and deteriorating production process. The system addresses the corrective maintenance, preventive maintenance, and allowable shortage. The quality loss of conforming product is considered and Taguchi's asymmetric quadratic quality loss function is used for evaluating the product quality. The optimal process mean and economic manufacturing quantity are jointly determined by minimizing the total expected cost of product per unit time. From the above numerical results, we have the following conclusion: the material cost and demand rate per unit time have a major effect on the total relevant cost of product per unit time. Further study should consider the unknown process standard deviation and address the joint design of manufacturing quantity, specification limits, process mean, and standard deviation.

References

[1] Porteus E L 1986 *Operations Research* **34** pp 137 − 144
[2] Rosenblatt M J and Lee H L 1986a *IIE Transactions* **17** pp 48 − 54
[3] Rosenblatt M J and Lee H L 1986b *IIE Transactions* **18** pp 2 − 9
[4] Dajmaludim V D, Murthy N P and Wilson R J 1994 *International Journal of Production*

Economics **33** pp 97 – 107

[5] Yeh R H and Lo H C 1998 *International Journal of Operations and Quantitative Management* **4** pp 265 – 275

[6] Yeh R H, Ho W T and Tseng S T 2000 *European Journal of Operational Research* **120** pp 575 – 582

[7] Wang C H and Sheu S H 2000 *Computer and Mathematics with Applications* **40** pp 1297 – 1314

[8] Wang C H and Sheu S H 2003 *European Journal of Operation Research* **49** pp 131 – 141

[9] Wang C H 2004 *Computers & Operations Research* **31** pp 2021 – 2035

[10] Wang C H 2006 *Naval Research Logistics* **53** pp 151 – 156

[11] Yeh R H and Chen T H 2006 *European Journal of Operational Research* **174** pp 766 – 776

[12] Mujahid S N and Rahim M A 2010 *International Journal of Operational Research* **9** pp 227 – 240

[13] RahimM A and Fareeduddin M 2011 *International Journal of Industrial and Systems Engineering* **8** pp 298 – 325

[14] Akram M A, Al-Saif A W and Rahim M A 2012 *International Journal of Industrial and Systems Engineering* **11** pp 375 – 405

[15] Sarkar B, Sana S S and Chaudhuri K 2010 *International Journal of Mathematics in Operational Researc*h **2** pp 467 – 490

[16] Sana S S, Goyal S K and Chaudhuri K 2007 *International Journal of Production Economics* **105** pp 64 – 80

[17] Sana S S 2010 An economic production lot size model in an imperfect production system *European Journal of Operational Research* **201** 158-170

[18] Lin G C and Gong D C 2018 A manufacturer-buyer model with random breakdowns *Proceeding of the Asia Pacific Industrial Engineering & Management Systems Conference* (Hong Kong)

[19] Springer C H 1951 *Industrial Quality Control* **8** pp 36 – 39

[20] Taguchi G 1986 *Introduction to Quality Engineering* (Tokyo: Asian Productivity Organization)

[21] Jeang A 2012 *Omega* **40** pp 774 – 781

[22] Chen C H and Tsai W R 2016 *Journal of Industrial and Production Engineering* **33** pp 495 – 500

[23] Chen C H, Chou C Y and Kan C C 2015 *Journal of Industrial and Production Engineering* **32** pp 196 – 203

[24] Chen C H, Khoo M B C, Chou C Y and Kan C C 2015 *Journal of Industrial and Production Engineering* **32** pp 219 – 224

Research on harmonic suppression ability of active power filter

X Shen[1], H C Shu[1] and M Cao[2]

[1]Mechanical and Electrical Engineering College, Kunming University of Science and Technology, China
[2]Yunnan Electric Power Research Institute, China

E-mail: 23755803@qq.com

Abstract. The system control model of double resonance injection hybrid active power filter (DIHAPF) and mathematical model of voltage source inverter with controller was established based on the structure and operating principle of DIHAPF. The injection branch features were analyzed including the static reactive power compensation capability and dividing capability of harmonic voltages. Stability and harmonic suppression capacity of DIHAPF was analyzed simultaneously. The influence of the parallel resonant circuit resonant frequency to system performance was discussed in detail. Furthermore, the method of designing the resonant frequency of parallel resonant circuit was proposed. The simulation model of DIHAPF was established and simulated on PSIM 6.0 taking nonlinear loads as harmonic current sources. A 6kV DIHAPF experiment prototype controlled by TMS320F28335 was built. Simulation and experimental results prove the validity of the analysis in this paper.

1. Introduction

With the development of power electronics technology, the proportion of non-linear load in the power grid is increasing day by day, and the harmonic pollution in the power grid is getting more and more serious [1-4]. This brings direct and indirect losses to power grid enterprises and power users, and also causes energy waste. In order to ensure the safe and stable operation of the power grid, protect the users' electrical equipment and save power and energy, it is imminent to control the harmonic of the power grid [5-7]. The active power filter is an important method of harmonic control. Among them, the injection hybrid active power filter (IHAPF) has the ability of harmonic suppression and large-capacity reactive power static compensation, taking into account the passive filter and the advantages of active filters [8-10]. In order to improve the application range of the hybrid active power filter and improve the compensation performance, a new double resonance injection branch is proposed in paper [11]. The characteristics of the double resonance injection hybrid active power filter (DIHAPF) and its application in power grid are discussed in paper [12]. DIHAPF has become the new harmonic control device for medium and high voltage power systems.

In the existing literature of DIHAPF characteristic analysis, the inverter is regarded as the voltage source or the current source for controlling the current control, and the output voltage or current is proportional to the control current. Based on this, the model is established and analyzed [13-14]. This model simplifies the influence of control links and inverters on the characteristics of the model and can not fully reflect the influence of the controller on the system characteristics. Therefore, the detailed DIHAPF model of the controller and its characteristics need further study.

Based on the analysis of structure and working principle of DIHAPF, a mathematical model of voltage source inverter containing controller is established, and then a DIHAPF mathematical model is established. The injection branch characteristics, system stability and harmonic compensation characteristics of DIHAPF are analyzed. The influence of resonant frequency of parallel resonant branch on the system performance is emphatically discussed, and the design method of resonant frequency is put forward. The results of simulation and experiment prove the correctness of this analysis.

2. DIHAPF model and working principle

Double resonance active power filter DIHAPF topology shown in Figure1.In the figure, C1, L1, C2, L2 constitute the injection branch. C1 and L1 form a series resonant branch, resonating at the fundamental frequency. C2, L2 constitute a parallel resonant branch.

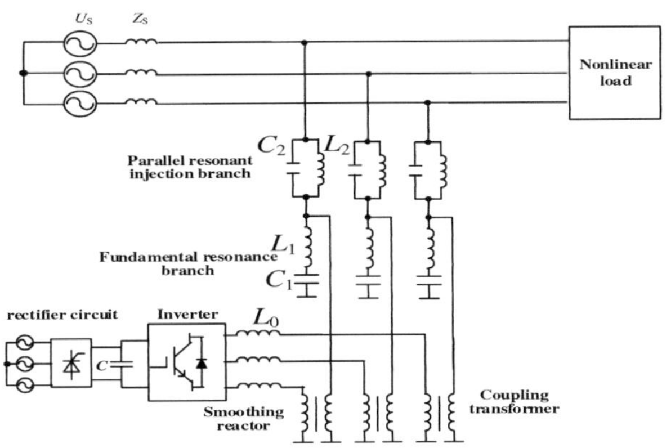

Figure 1. Topology of DIHAPF

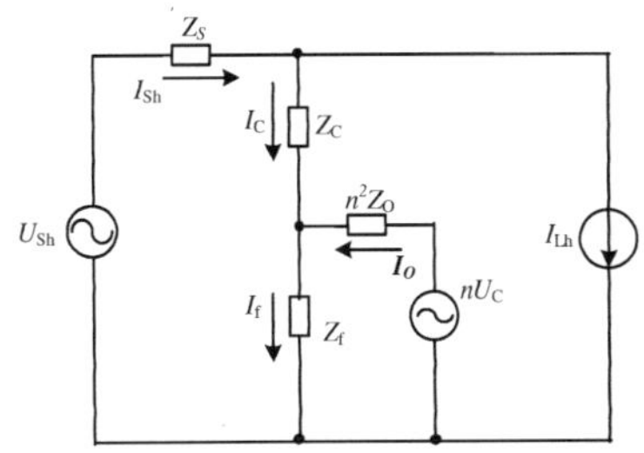

Figure 2. Single-phase equivalent circuit

By simplifying the DIHAPF, treating the non-linear load as a harmonic current source and controlling the active part of the DIHAPF as a voltage source, the single-phase equivalent circuit of The single-phase equivalent circuit of double resonance injection hybrid systemin the harmonic domain is shown in Figure 2. USh is the harmonic voltage of the power grid, ZS is the harmonic impedance of the power grid, ILh is the harmonic current of the load, ZC is the injection impedance, Zf is the impedance of the fundamental resonance branch, and ZO is the impedance of the smoothing reactor.

According to the equivalent circuit diagram, the following equations can be obtained from Kirchhoff's voltage and current law.

$$U_{sh} = Z_s I_{sh} + Z_c I_c + Z_f I_f \tag{1}$$

$$nU_c = n^2 Z_o I_o + Z_f I_f \tag{2}$$

$$I_{sh} - I_c - I_{Lh} = 0 \tag{3}$$

$$I_c + I_o - I_f = 0 \tag{4}$$

Solve the simultaneous equations above, eliminating Ic, I1, I2 and Ip. According to load harmonic current control mode, the active part output voltage is controlled to K times of load harmonic voltage, and then

$$I_{sh} = \frac{(n^2 Z_0 + Z_f) U_{sh} + k_1 I_{Lh}}{n^2 Z_0 (Z_c + Z_s) + Z_f (Z_c + Z_s + n^2 Z_0)} \tag{5}$$

Where $k_1 = n^2 Z_0 Z_f + n^2 Z_0 Z_c + Z_c Z_f - nKZ_f$

When the DIHAPF active part output voltage is considered only, the system equivalent circuit is shown in Figure3.

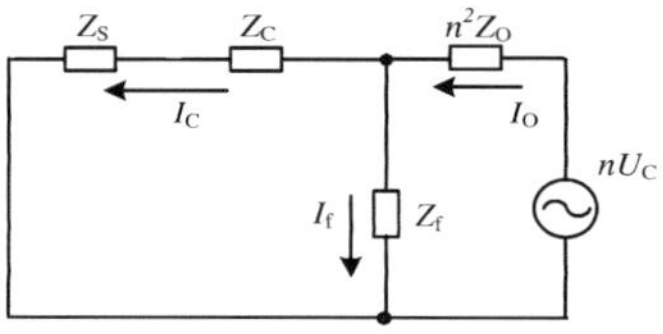

Figure 3. Single-phase equivalent circuit taking the output voltage of active part as the only source

From Figure 3, the following equation is obtained.

$$I_c = \frac{nZ_f}{(Z_s + Z_c) Z_f + n^2 Z_o (Z_s + Z_c + Z_f)} U_c \tag{6}$$

Considering the voltage source inverter, and in accordance with the load harmonic current control mode, DIHAPF system control model shown in Figure4.

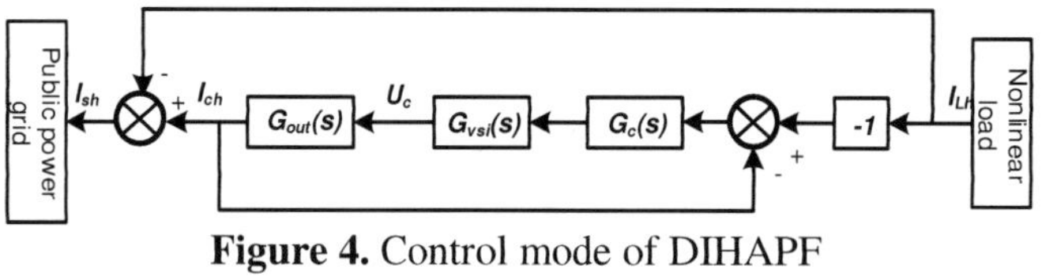

Figure 4. Control mode of DIHAPF

Assuming that the transfer function of the inverter output voltage Uc and the harmonic current Ic injected into the power grid by the active filter is Gout, and then

$$G_{out} = \frac{nZ_f}{(Z_s + Z_c) Z_f + n^2 Z_o (Z_s + Z_c + Z_f)} \tag{7}$$

As seen in Figure 4.

$$U_c = G_{vsi}(s) G_c(s)(-I_{Lh} - I_{ch}) \tag{8}$$

$$I_{ch} = G_{out}(s) U_c \tag{9}$$

The mathematical model of the voltage-source inverter is derived from Eqs. (8) and (9).

$$U_{\mathrm{c}} = -\frac{G_{\mathrm{vsi}}(s)G_{\mathrm{c}}(s)}{1+G_{\mathrm{out}}(s)G_{\mathrm{vsi}}(s)G_{\mathrm{c}}(s)} I_{\mathrm{Lh}} \tag{10}$$

3. Analysis of DIHAPF characteristics

3.1. Injection branch characteristics

The injection branch consists of an injection section and a fundamental series resonance circuit (FSRC). The active part of the active filter is connected in parallel with the fundamental series resonant branch via the coupling transformer. Therefore, the capacity of the active part is determined by the voltage division capacity of the fundamental series resonant branch and the output current capacity of the inverter. The partial pressure of the fundamental series resonant branch is mainly composed of two parts, which are the partial pressure of the power grid on the FSRC and the harmonic voltage of the injection branch near the tuning frequency.

The impedance of the IHAPF injection branch can be expressed as

$$Z_1 = R_{11} + \mathrm{j}(\omega L_{11} - \frac{1}{\omega C_{11}} - \frac{1}{\omega C_{12}}) \tag{11}$$

Double resonance injection APF injection branch impedance can be expressed as

$$Z_2 = [R_{21} + \frac{R_{22}}{k_2}] + \mathrm{j}[\omega L_{21} - \frac{1}{\omega C_{21}} + k_3] \tag{12}$$

Where $k_2 = (1-\omega^2 L_{22}C_{22})^2 - \omega^2 C_{22}{}^2 R_{22}{}^2$ and $k_3 = \frac{1-\omega^2 L_{22}C_{22} - \omega C_{22}R_{22}{}^2}{(1-\omega^2 L_{22}C_{22})^2 - \omega^2 C_{22}{}^2 R_{22}{}^2}$.

The Matlab7.0 simulation software is used to simulate the IHAPF injection structure and the DIHAPF injection structure. The system simulation parameters are shown in Table 1.

Table 1. Simulation parameters of injection branch.

Parameter	IHAPF	DIHAPF
Reactance of fundamental series resonant branch	14.58 mH	14.58 mH
Capacitance of the fundamental series resonant branch	690 uF	690 uF
Internal resistance of reactance of fundamental series resonant branch	0.1 ohm	0.1 ohm
Capacitance of the injection part	36 uF	50 uF
Reactance of the injection part	--	167.47 mH
Internal resistance of reactance of injection part	--	0.1 ohm

The impedance characteristics of the two structures and the frequency amplification of the series resonant branch voltage are analyzed respectively. The simulation results are shown in Figure 5.

3.2. Analysis of reactive power and static power

It can be seen from Figure 5(a) that the impedance of the IHAPF injection branch is capacitive and gradually decreases until it reaches zero when the frequency is less than the resonance frequency. When the frequency is greater than the resonance frequency, the impedance of the injection branch is inductive and gradually increasing. The impedance of DIHAPF injection branch changes suddenly at the resonance point. When the frequency is lower than the resonance frequency, the impedance is capacitive and gradually increasing. When the frequency is greater than the resonance frequency, the impedance is inductive and gradually decreasing. At the fundamental frequency, the injection branch impedances of IHAPF and DIHAPF are both capacitive and have reactive static compensation capability. At each frequency point after the resonance point, the DIHAPF injection branch impedance is greater than that of IHAPF. Because the impedance characteristics of the fundamental series resonant branch are the same, the partial pressure characteristic of the DIHAPF fundamental series resonant branch is better, which effectively reduces the withstand voltage of the active part. At the same time, due to the existence of the parallel inductance, which can control the reactive power of the parallel branch injection system and avoid the reactive power compensation of the system.

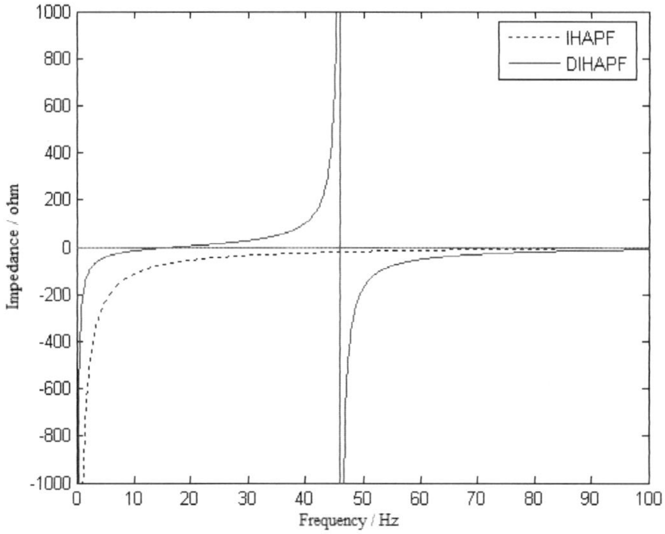

(a)

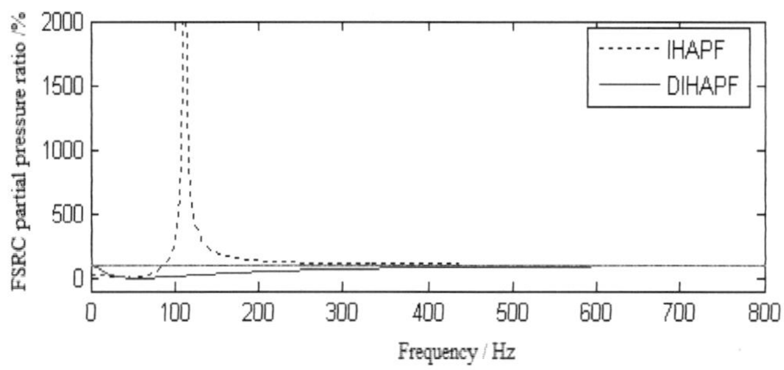

(b)

Figure 5. Impedance characteristics of injection branch:(a) injection branch impedance and (b) partial pressure ratio of fundamental wave resonance branch

3.3. Analysis of harmonic partial pressure characteristics

The voltage division capability of the fundamental series resonant branch affects the active capacity of the hybrid active power filter. The harmonic partial pressure characteristics of the FRSC are analyzed by the ratio of the impedance of the fundamental series resonant branch and the total impedance of the injected branch. The fundamental and harmonic partial pressure data of the two kinds of implanted structural fundamental wave series resonant branches are shown in Table 2.

Table 2. The voltage ratio of fundamental series resonance branch(%)

	IHAPF	DIHAPF
Fundamental	0.20	0.04
3th	200.16	31.84
5th	120.01	61.68
7th	109.09	76.93
11th	103.45	89.48
13th	102.44	92.28

From Figure 1(b) and Table 2, it is known that the fundamental voltage of single implanted APF injection branch is 0.2%, however, the harmonic voltage of the second order and above is amplified and larger than the corresponding harmonic voltage of the power grid. The fundamental voltage of double resonance injection APF series resonant branch is 0.04%, and the harmonic voltages are lower than the corresponding harmonic voltage of the power grid,the operating voltage of the active part is reduced, and then the active capacity of the DIHAPF is reduced.

3.4. DIHAPF stability and harmonic rejection characteristics

Ignoring the harmonic voltage of the power grid, only considering the influence of the load harmonic current, the harmonic suppression function of DIHAPF can be obtained as

$$G_{\mathrm{h}}=\frac{I_{\mathrm{sh}}}{I_{\mathrm{Lh}}}=\frac{n^2Z_0Z_{\mathrm{f}}+n^2Z_0Z_{\mathrm{c}}+Z_{\mathrm{c}}Z_{\mathrm{f}}-nKZ_{\mathrm{f}}}{n^2Z_0(Z_{\mathrm{c}}+Z_{\mathrm{s}})+Z_{\mathrm{f}}(Z_{\mathrm{c}}+Z_{\mathrm{s}}+n^2Z_0)} \tag{13}$$

For convenience of discussion, the effect of system delay on stability is not considered, the open-loop transfer function of DIHAPF system is

$$G_{\mathrm{o}}=\frac{[n^2Z_{\mathrm{o}}(Z_{\mathrm{c}}+Z_{\mathrm{f}})-Z_{\mathrm{c}}Z_{\mathrm{f}}-nKZ_{\mathrm{f}}]Z_{\mathrm{p}}}{n^2Z_{\mathrm{o}}k_4-Z_{\mathrm{c}}Z_{\mathrm{f}}(Z_{\mathrm{s}}+Z_{\mathrm{p}})} \tag{14}$$

Where $k_4=(Z_{\mathrm{c}}+Z_{\mathrm{p}})Z_{\mathrm{f}}+(Z_{\mathrm{c}}+Z_{\mathrm{f}}+Z_{\mathrm{p}})Z_{\mathrm{s}}$

Suppose the system characteristic equation is

$$[b_1\ \ b_2\ \ \ldots\ \ b_8][s^7\ \ s^6\ \ \ldots\ \ s^0]^{\mathrm{T}}=0 \tag{15}$$

The simplified DIHAPF system is a linear, stationary system. The necessary and sufficient condition for the stability of a linear stationary system is that the eigenvalues of the system lie in the left half of the complex plane. Therefore, whether the DIHAPF system is stable or not is judged according to whether the eigenvalues of the open-loop transfer function of the DIHAPF system are all located in the left half plane of the s domain. DIHAPF described in this article is a high-level system, the stability of the system can be based on the symbols of the first row of the Rolls-table parameter to determine system stability.

When the resonant frequency of parallel branch is 42 Hz, 44 Hz and 48 Hz respectively, the value of the first column of the Routh table of the characteristic equation of the DIHAPF system is shown in Table 3.

According to the Routh criterion, it is known that the DIHAPF system is stable at the three resonant frequencies.

When the frequency is less than the resonant frequency, the total impedance of the parallel resonant circuit is inductive; greater than the resonant frequency, was capacitive. DIHAPF has a certain capacity of capacitive reactive power static compensation, and the resonant frequency of the parallel resonant branch should be less than the fundamental frequency.

Table 3. Rolls table under different resonance frequencies

	Fr=42Hz	Fr=44Hz	Fr=48Hz
s^7	8.262e-23	8.262e-23	8.262e-23
s^6	8.265e-18	8.265e-18	8.265e-18
s^5	2.715e-16	2.724e-16	2.715e-16
s^4	1.108e-11	8.113e-11	2.594e-11
s^3	1.146e-9	8.421e-10	2.782e-10
s^2	1.348e-5	9.814e-6	3.006e-6
s^1	2.1e-6	2.1e-6	2.1e-6
s^0	2.1e-6	2.1e-6	2.1e-6

The static reactive power compensation ability of the injected branch is guaranteed, and simultaneously, the influence of the resonant frequency of the parallel resonant branch on the system stability is analyzed. The parallel branch parameters are calculated at the resonant frequency of 42Hz, 44Hz and 48Hz respectively. The open loop transfer function of the system is analyzed by Matlab, and the Potter diagram is shown in Figure 6.

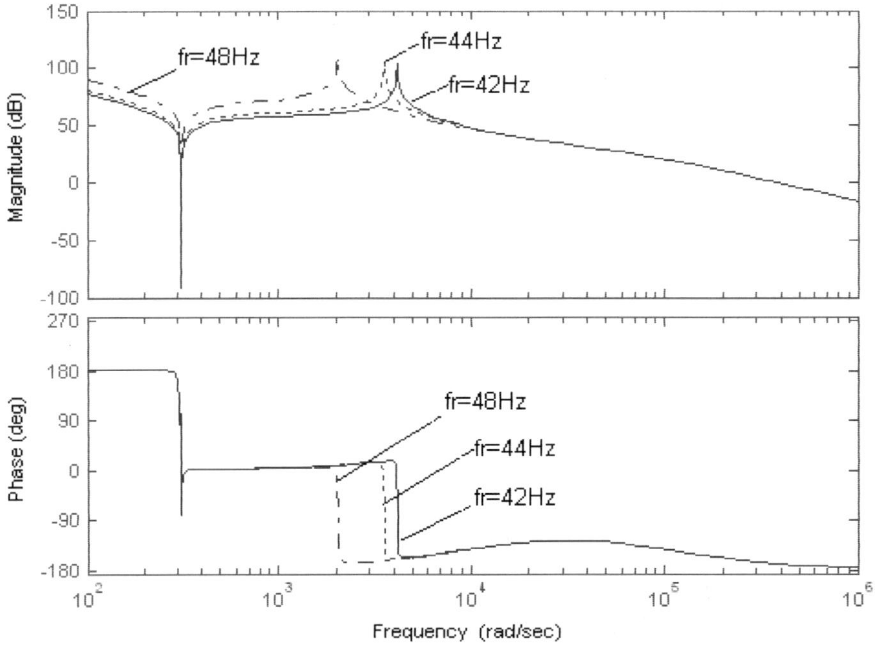

Figure 6. Open-loop Bode plot of DIHAPF as fr=42 Hz, 44 Hz, 48 Hz

It can be seen from Figure 6 that the phase angle curve of the DIHAPF system has three points of abrupt change. The first point of abrupt change is caused by the parallel resonant branch, the frequency is 299 rad / sec, the second point of abrupt change is caused by the fundamental series resonance branch. The third mutation frequency is the resonant frequency of the injection branch. When the resonant frequency of the parallel branch is 42 Hz, 44 Hz and 48 Hz respectively, the resonant frequency of the injected branch is 4 260 rad/sec, 3 680 rad/sec and 2 070 rad/sec. After the phase angle is stable, the phase-frequency curve gradually increases and gradually approaches -180°. Maintaining the other parameters unchanged and guaranteeing the static reactive power compensation ability of the fundamental, simultaneously, when the resonance frequency of parallel branch is 42Hz, 44Hz and 48Hz, respectively, the phase margin of open-loop transfer function of DIHAPF system is 107.93° 106.21° and 103.44°, respectively. As a result, when the resonant frequency is larger, the stability of DIHAPF is poor. With the decrease of the resonant frequency of the parallel branch, the system angle margin increases and the stability is improved.

The influence of the resonant frequency of the parallel resonant branch on the harmonic suppression ability is analyzed. While keeping the other parameters unchanged, as well as guaranteeing the static reactive power compensation ability of the fundamental,the parameters of the parallel resonant branch are adjusted to resonate at 42Hz, 44Hz and 48Hz, respectively, and the influence on the harmonic suppression capability of DIHAPF is shown in Figure 7.

It can be seen from Figure 7 that the gain of DIHAPF in the main harmonic frequency band is approximately -58.4dB, -62.7dB and -73.3dB, respectively, when the resonance frequency of parallel resonant branch is 42Hz, 44Hz and 48Hz, respectively. It shows that the harmonic suppression ability of DIHAPF system is poor when the resonance frequency is small, and the harmonic suppression ability of DIHAPF system increases with the increase of resonance frequency.

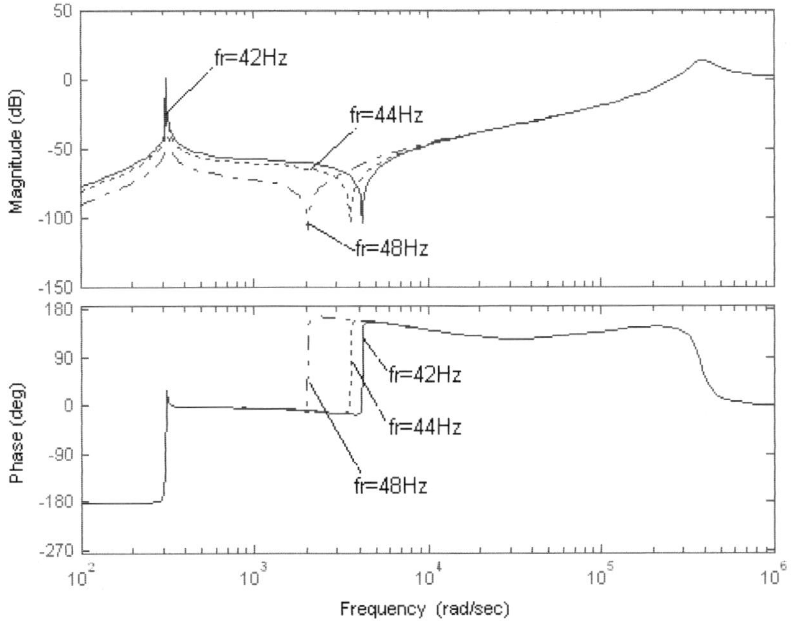

Figure 7. Harmonic suppression characteristic of DIHAPF as fr=42Hz, 44Hz, 48Hz

3.5. Selection of resonant frequency of parallel branch

By analyzing the stability and harmonic filtering ability of DIHAPF, the resonant frequency of parallel branch must be less than the fundamental frequency to ensure the static reactive power compensation capability of DIHAPF system. When the harmonic frequency is less than the fundamental frequency, the resonant frequency is smaller. The better the system stability, the worse the harmonic suppression; the greater the resonance frequency, the better the system harmonic suppression, and the worse the system stability. Therefore, when designing the DIHAPF system parameters, the system stability and harmonic control effects need to be considered synthetically. The proper resonant frequency should be chosen so that the DIHAPF system can balance the stability and harmonic control effects.

In this paper, the resonant frequency of 44Hz is selected. Under this parameter, the phase margin of DIHAPF system is 106.21°, and the gain of the main harmonic frequency is about -62.7dB.

4. Simulation and experimental research

In order to verify the correctness of the above analysis and the comprehensive compensation effect of DIHAPF structure, according to the structure and principle of DIHAPF, the software PSIM6.0 is used for simulation. The harmonic is generated by the equivalent nonlinear load of the harmonic current source [15]. The DIHAPF system parameters are determined according to the harmonic control and reactive power compensation requirements of a steel plant. The simulation model is powered by three-phase AC power supply. The power grid model consists of voltage source, grid impedance, transformer and transmission line. The load consists of load impedance and random harmonic current sources. Among them, the grid parameters are as follows: the three-phase power line voltage is 35kV, the frequency is 50Hz, the equivalent impedance of the grid is 0.05mH and the internal resistance is 0.001Ω. The harmonic source parameters are as follows: the 5th harmonic current is 28A, the 7th harmonic current is 13.6A, the 11th harmonic current is 8.3A, the 13th harmonic current is 6 A, and the 17th harmonic current is 3.7A . The DHIAPF parameters are as follows: output inductance 0.2mH, internal resistance 0.005Ω; fundamental series resonant branch inductance 14.58mH, internal resistance 0.1Ω, capacitance 690uF; parallel resonant branch inductance 109.03mH, internal resistance 0.1Ω, capacitance 120uF; coupling transformer voltage ratio of 2:1; inverter by the three-phase rectifier bridge power supply. Simulation results are shown in Figure 8, where IL is the load current; Is the grid current; Ich is the current injected into the grid by DIHAPF; Uc is the inverter output voltage.

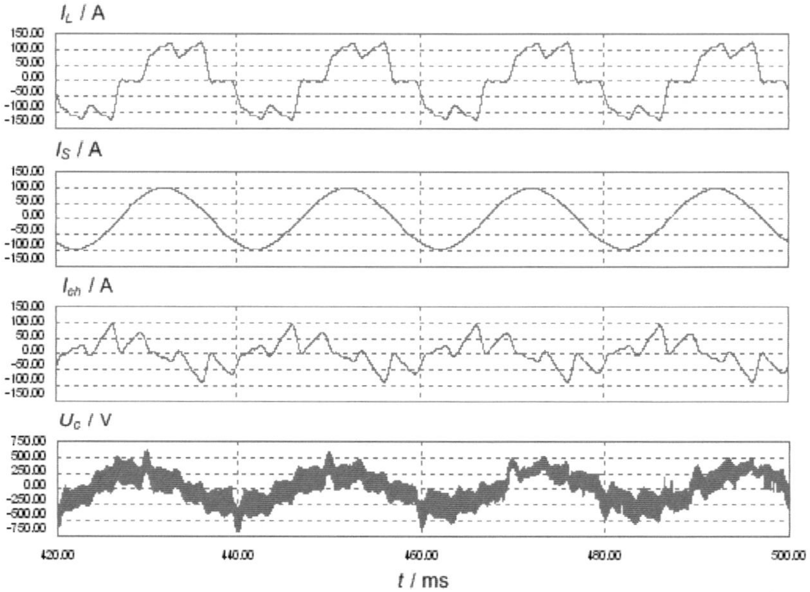

(a)

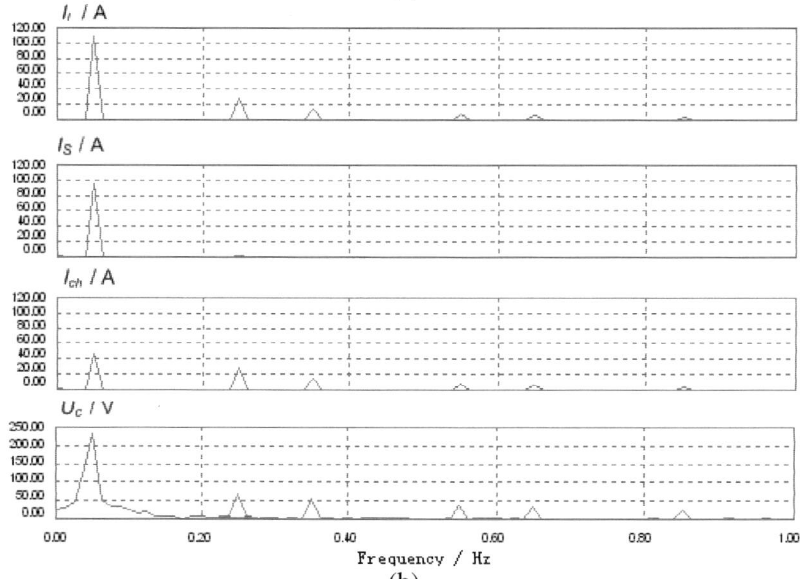

(b)

Figure 8. Simulation result:(a) simulation waveforms and (b) spectrum

Table 4. Comparison of harmonic current amplitude

The order of harmonic current	Before put into operation/A	After put into operation/A
5	27.52	1.21
7	13.51	0.58
11	8.17	0.48
13	5.95	0.44
17	3.54	0.34

From the simulation Figure 8, it can be seen that the harmonic current of the load flows directly into the power grid before DIHAPF input, causing serious harmonic pollution. After the operation of DIHAPF, the harmonic currents in the power grid are greatly reduced, and the harmonic pollution is effectively controlled. Harmonic current data before and after system input is shown in Table 4.

In order to further verify the characteristics of the dual-resonant injection hybrid active power filter, an experimental prototype for 6kV distribution network was developed in the laboratory. The active part of DIHAPF selects Mitsubishi IGBT-IPM series module PM300CLA120 to form a three-phase inverter bridge. The DC side capacitor is composed of four 10000uF electrolytic capacitors by two in series and two in series. System controller using TMS320F28335 digital DSP control program. The output filter is installed to filter the high-frequency switching harmonics generated by the inverter, the inductance of 0.2mH is selected in this paper. The fundamental series resonant branch has an inductance of 14.57mH and a capacitance of 690uF. The parallel resonant branch inductance is 25.31mH with a capacitance of 400uF. The ratio of the coupling transformer is 2. The oscilloscope is used to obtain the current waveform before and after the DIHAPF prototype is put into operation, as shown in Figure 9.

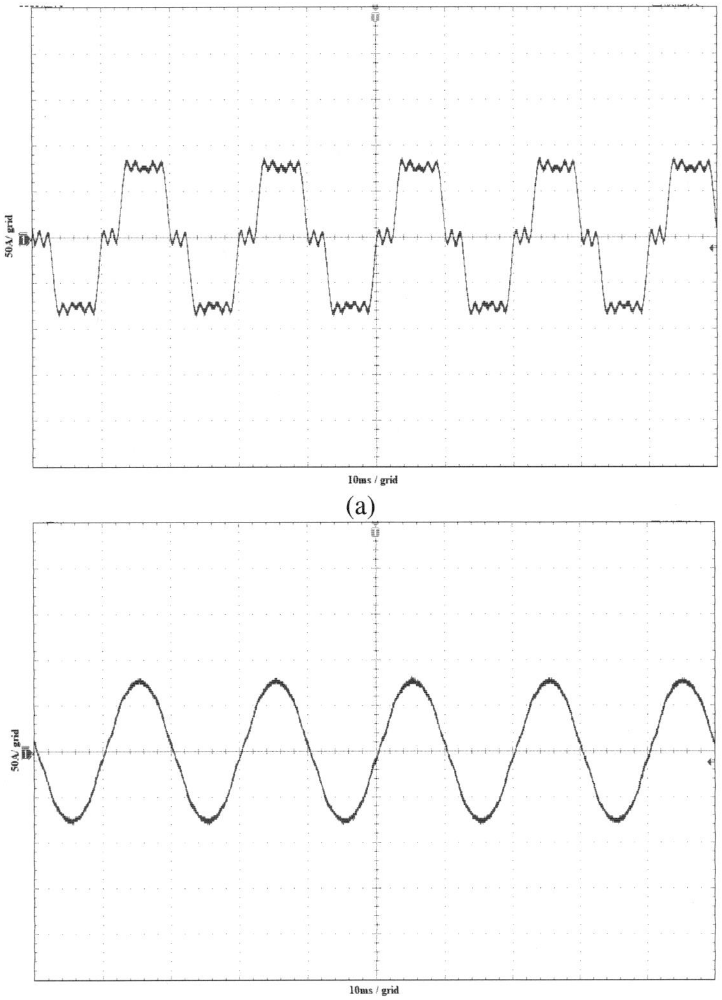

(b)

Figure 9. Waveform of grid current before and after operating prototype: (a) the Phase A phase current waveform of the grid before the DIHAPF unit is put into operation, and (b) the Phase A phase current waveform of the grid after the DIHAPF unit is put into operation

5. Conclusion

Through the analysis of DIHAPF characteristics and simulation experiments, we get the following conclusions:

(i) At the fundamental frequency, the impedance of DIHAPF injection branch is greater than the impedance of IHAPF injection branch. The fundamental voltage series resonant branch voltage dividing characteristics are good and the active part withstand voltage is reduced effectively, so that DIHAPF active capacity is smaller, and suitable for higher voltage levels. Moreover, shunt reactor can control the system into the reactive power to avoid reactive over-compensation.

(ii) When the resonant frequency of the parallel branch is determined, the stability of the given system and harmonic suppression become contradictory. In the design of DIHAPF system parameters, this paper presents a resonant frequency design method considering system stability and harmonic control effects.

(iii) Simulation and experimental results show that the DIHAPF system has good stability and harmonic governance ability, and verifies the correctness of the analysis.

References

[1] Zhu J et al 2014 *High Voltage Apparatus* **50**(5) pp 56 — 62
[2] Yang Q X et al 2012 *Power System Protection and Control* **40**(3) pp 119 — 123
[3] Yao Z Q et al 2013 *Proceedings of the* CSEE **33**(36) pp 27 — 33
[4] D ai K et al 2013 *Transactions of china electrotechnical society* **28**(9) pp 79 — 85
[5] Han Y et al 2014 *Proceedings of the CSU-EPSA* **26**(9) pp 1 — 6
[6] Yao Z Q, Zhao Q and Liu X M 2011 *Power System Protection and Control* **41**(24) pp 71 — 76
[7] Meng J L et al 2012 *Power System Protection and Control* **41**(14) pp 73 — 80
[8] Dirk D et al 2001 *IEEE Trans. on Power Electronics* **16**(6) pp 821 — 827
[9] Xu N, Xu H Y and Huang H H 2014 *Proceedings of the CSU-EPSA* **26**(3) pp 57 — 60
[10] Shen L F et al 2013 *Power System Protection and Control* **41**(4) pp 91 — 96
[11] Wang X W et al 2010 *Automation of Electric Power System* **34**(2) pp 59 — 62
[12] Deng X, Luo A and Zhao W 2011 *Power System Technology* **35**(10) pp 82 — 87
[13] Shuai Z K et al 2009 *Transactions of China Electrotechnical Society* **24**(5) pp 121 — 132
[14] Gou X K et al 2009 *Power System Protection and Control* **37**(17) pp 13 — 16
[15] Xia X Y, Luo A and Luo S W 2010 *Electric Power Automation Equipment* **24**(3) pp 195 — 202

A review of reconstruction research for the lost gearing of antikythera astronomical calculator

Jian Liang Lin

National Science and Technology Museum, Taiwan

E-mail: golduser007@gmail.com

Abstract. This paper presented a review for the reconstruction research of the Antikythera mechanism, the mechanical calculator was made around the end of second century BC in ancient Greece. Under the concluded design specifications and constraints, all possible conceptual designs of the lost mechanical structures subject to the standard of technology and science at that time were synthesized based on a systematic design procedure. This oldest astronomical calculator excavated in 1900-1901 can display the solar, lunar and planetary motions, indicate the date of Egyptian calendar, calculate the luni-solar calendars and periods, and predict the eclipses by its interior gear trains. The structures for the solar and the planetary motions are completely lost in the excavation. Without surviving evidence, restoring the lost structures by computer tomography or X-ray detection is unfeasible. Here, the author applied a design procedure with four steps, which is based on the reconstruction design methodology of lost ancient machinery. Historical archives for reconstructing Antikythera mechanism are built based on historical records, ancient astronomy, ancient astronomical instruments, as well as the kinematic and mechanism analyses of gear trains. After identifying input conditions, three and two design specifications respectively for the solar and the planetary motions are concluded and their feasible designs are synthesized by the design procedure. For each design, the equations of teeth counting are derived based on the analysis of gear ratios. Finally, the possible teeth of gears are generated and then a new reconstruction design of Antikythera mechanism with complete interior structure is built through the constraints of space and assembly. These synthesized designs form a database of reconstruction work, and one of them can be the original structure. Once new evidence is discovered, these reconstruction designs can help archaeologists decode this mechanism further.

1. Introduction

Antikythera mechanism is the oldest mechanical calculator for astronomical applications that includes demonstrations of heavenly bodies' motions, calculations of calendars, predictions of eclipses, and display of cyclic moon phase. The mechanism was excavated from the sea area of Antikythera Island in 1900-01, and its excavation, a group of fragments, are preserved in national archaeological museum of Aegean. The mechanism is suggested as a wooden box approximately 315x 190x 100 mm which covers complicated designs of gear trains containing bronze gears with triangular teeth shape [1-2] Figure 1(a) is the reconstruction of known appearance [3].

Several opinions, reconstruction designs and physical models about the Antikythera device have been presented after 1901 [1, 2, 4-28]. From 1902 to 1974, some scholars, such as Statis, Svoronos, Rediadis, and Rehm etc., made research to define the category of the device. In 1974, Price presented the first systematic reconstruction designs with differential gearing to describe the motions of the Sun and the Moon on the front plate and the eclipses on the lower part of back plate. From 2002 to now, Wright made a series research to present the second systematic reconstruction design. Metonic and Callippic periods were indicated on the back upper dials and the corresponding gearing were reconstructed. An epicyclic gearing with a pin-in-slot joint was introduced to generate the lunar motion. Freeth et al. decoded a large number of inscriptions to confirm the dials' functions and reconstruct the third systematic design. The design of spiral dials, Olympiad period, Saros period and Exeligmos period were presented and the epicyclic gear train with a pin-in-slot joint was revised to generate the lunar anomaly motions. And, the possible lost gearing was reconstructed in 2012. Furthermore, Koetsier introduced the historical development of reconstruction work in2008. Evan et al. presented the possible gearing and the front dial design for the lunar and solar motion. Also, they presented the analysis of pin-in-slot device. The previous studies show several continuous research results for the possible reconstruction designs. The functions and the systematic interior structure was were gradually defined. Focusing on the lost gearing, the topic to demonstrate the solar motion and planetary motions were confirmed. The pin-in-slot device should be applied in the lost gearing. However, the reconstruction designs by a systematic procedure base on the concepts of mechanisms are not indicated either Freeth et al. and Evans et al.

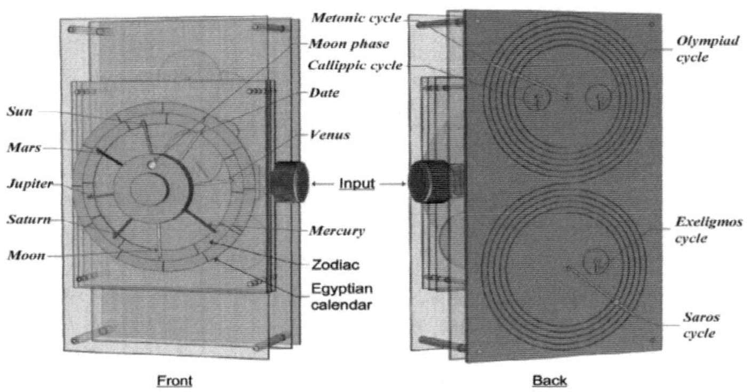

(a) appearance and functions []

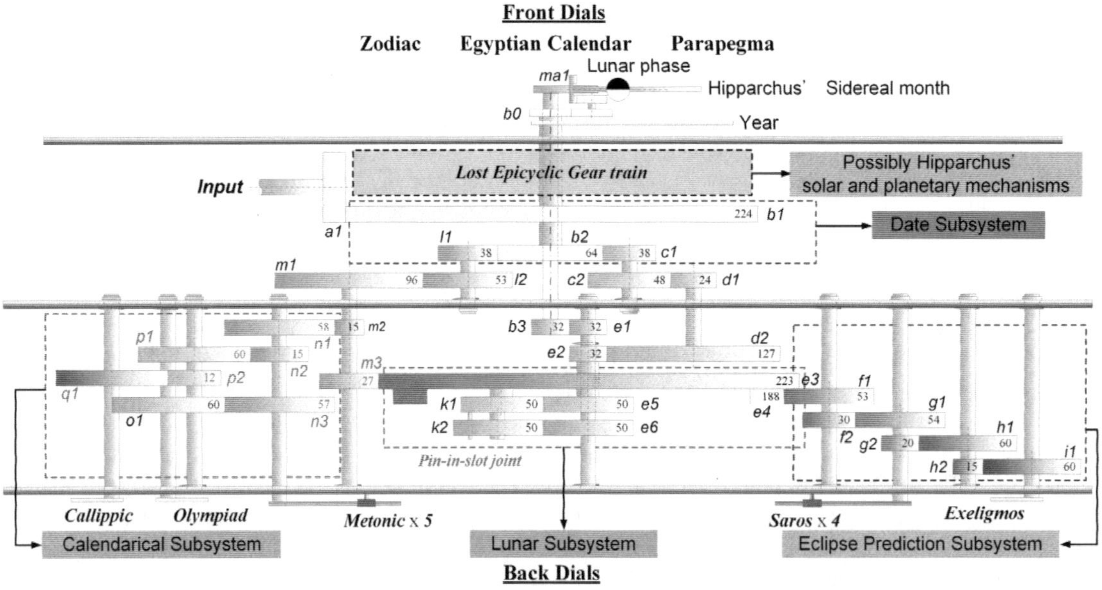

(b) a reconstruction design of interior gearing by Freeth et. Al [1-2]

Figure 1. The Knwon Antikythera Astronomical Device

IOP Conf. Series: Materials Science and Engineering **658** (2019) 012014 doi:10.1088/1757-899X/658/1/012014

In accordance with the viewpoint of mechanical structure and the assigned functions, the reconstruction design of Antikythera device that is the most accepted by academia can involve six subsystems, as shown in Fig. 1(b) [1-2]. They are the date subsystem, the lunar subsystem, the calendrical calculation subsystem, the eclipse prediction subsystem, the solar subsystem and the planetary subsystem. The date subsystem, with complete gearing and sufficient surviving evidence, is to generate the daily motion then indicate on the outer of the front dials for the Egyptian calendar. The eclipse prediction subsystem, with complete gearing and sufficient surviving evidence, is to generate the Saros cycle and the Exeligmos cycle respectively on the lower 5-turns spiral dial and the circle dial of back plate. The lunar subsystem, with an unclear feature about coaxial arbors in surviving evidence, displays the lunar anomaly motion based on Hipparchus' lunar theory. The surviving evidence of calendrical calculation subsystem is incomplete and at the upper of back plate. The subsystem shows the Metonic cycle on the 4-turns spiral dial, the Callippic cycle and the Olympiad cycle on two circle dials respectively. The reconstructions of above subsystems are mostly supported by the excavation of Antikythera mechanism. However, the gearing, which corresponds to the solar subsystem demonstrating the motion of Sun and the planetary subsystem demonstrating the motions of Mercury, Venus, Mars, Jupiter and Saturn, are almost completely lost, except for the remaining features of arbors on the surviving gear.

Since 2007, the author and Prof. Yan made efforts to the reconstruction research of Antikythera device and presented a series of results [3, 29-33]. The purpose of paper is to review the reconstruction research of the lost gearing for the solar and planetary motions in the past years, and present the interpretation of the synthesized designs according to the finding of surviving evidence.

2. Design procedure

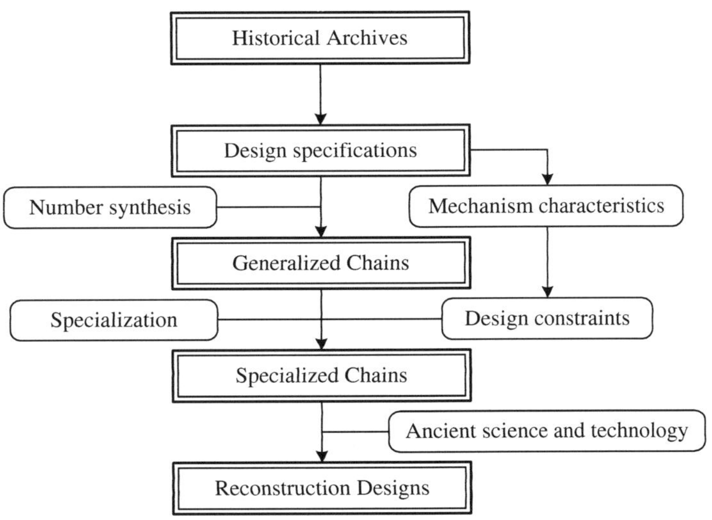

Figure 2. Procedure of reconstruction synthesis.

Undoubtedly, the reconstruction designs must be consistent with the level of science, technology and manufacturing techniques at the time period of Antikythera device. In order to reconstruct the lost solar and planetary subsystems, this reconstruction research applies a systematic design procedure, as shown in Fig. 2 [3]. The procedure includes five steps: historical archives, design specifications, generalized chains, specialized chains and reconstruction designs. In fact, the historical archives of design procedure are varied with different research topics, and the applications of design procedure are flexible. This design procedure, which is defined based on the concepts of mechanism number synthesis, is contributed to restore other lost ancient machines [34-36]. Following the procedure, the possible conceptual designs of each subsystem are synthesized. However, the conceptual designs of subsystems are generated respectively. The constraints of space and the interference of subsystems (lost and existing structures) should be considered further during the assembly work for a complete interior structure.

2.1. Historical Archives

The research scope of the historical archives for the reconstruction design of Antikythera mechanism are defined as follows: the surviving evidence, the decoded information, the ancient astronomy, the ancient astronomical instruments, and the kinematic and mechanism analyses. It is known that the solar and the planetary motions are demonstrated on the front concentric dial through the investigation of the initial four works. The kinematic characteristics of the solar and the planetary anomaly motions are concluded from the ancient astronomical theories. Moreover, the anomaly motions of celestial bodies are demonstrated completely by the gear mechanisms with a pin-in-slot joint. This mechanism design is proved in several ancient astronomical instruments and the kind of designs and techniques can be retained. Furthermore, three possible input conditions serving for the lost subsystems are suggested by understanding the existing reconstruction designs, as shown in Fig. 3 [3, 30, 31]. In these input conditions, the input links provide different rotation directions. Also, the teeth of input link can be calculated to generate the required input rate by the analysis of gear ratios.

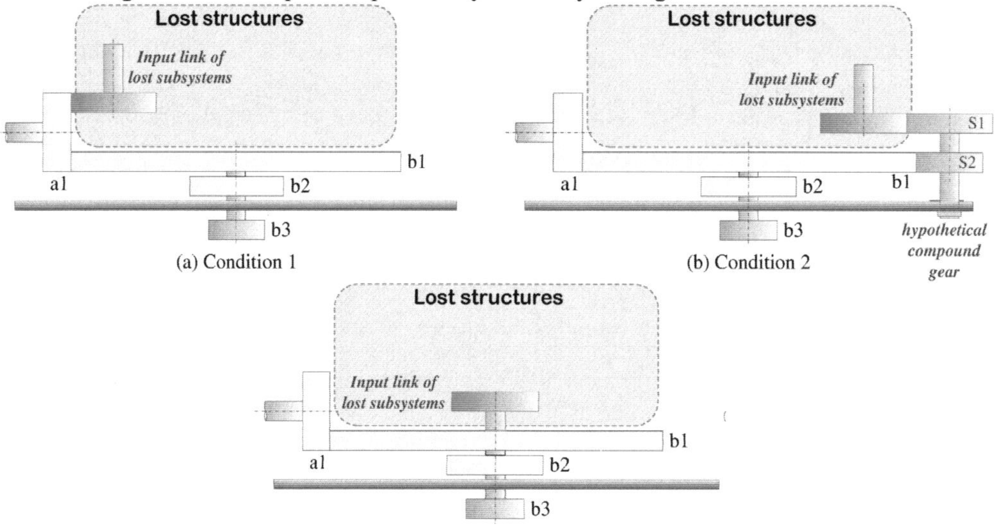

Figure 3. Feasible input conditions of the lost subsystems.

2.1.1. Surviving evidence. The computer tomography or X-ray detection technology in modern time is applied for detecting surviving fragments of device. With the advanced technology, the existing gears covered in the corroded fragments are clearly discovered. Therefore, the size and shape of gears, the teeth, the number of elements, the category of elements (pin-in-slot joint) as well as the features of arbors (like the pentagon feature on the epicyclic gear train) can be identified. The database of detection should be satisfied for the reconstruction work. Also, the surviving evidence is the reference for reconstructing the gears of lost gearing.

2.1.2. Decoded information. The large number of inscriptions, glyphs and symbols over the appearance of fragments are identified during the process of detection. For the past more than 100 years, several archaeologists and historians endeavoured to decode these inscriptions, glyphs and symbols. The functions of device are realized with the decoding of the inscriptions of dials. The periods demonstrated on the dials are the output rates of interior subsystems. Therefore, they are the important for reconstructing the gearing and the corresponding teeth.

2.1.3. Ancient astronomy. Antikythera astronomical device dates about the time period from the late second century BC to the early first century BC. The sophisticated design and complicated mechanical structure of device obviously represent the amazing achievements on astronomy at that time. Undoubtedly, this mechanism is built by following the observation records of astronomical phenomena and the astronomical theories. Therefore, the study of astronomy subject to the time period is indispensable. Especially for the lost gearing, the reconstruction designs should follow the investigations of ancient astronomical theories, i.e., the solar and planetary subsystems should be the demonstration of epicycle system (epicycle-deferent) by Apollonius and Hipparchus.

2.1.4. Ancient astronomical instruments. The techniques, theories, knowledge can be inherited and improved. In the historical development, the amazing mechanical devices often appeared in different time periods and areas. The same designs and techniques are applied in the same kind of devices. Focusing on the astronomical instruments that demonstrated the astronomical theories, the manufacture techniques are regarded as a traditional group of the craftsmen. Logically, not focusing on Greek ones, other ancient astronomical instruments could apply the same designs to achieve the required functions, even though none of them is Antikythera mechanism. Therefore, the study of ancient astronomical instruments, including Islamic astrolabe with gearing, Byzantine astrolabe with gearing, Persian astrolabe with gearing and Dondi's astronomical clock, is useful to the reconstruction study of lost gearing

2.1.5. Kinematic and mechanism analysis. For the lost subsystems, this work is indispensable to conclude the numbers of links and joints, the types of mechanisms and mechanical elements. It is understood that the solar and the planetary motions are anomaly and periodic based on the ancient astronomy. The application of pin-in-slot joint can be used to generate such an anomaly motion by the experience of reconstructing the lunar subsystem [32]. The gear train is applied to simulate the epicycles system (the deferent and the epicycle of astronomical model) and the eccentric system. A mechanism with fewer members to is better for the same purpose or functions. As a result, gear trains with a pin-in-slot joints are determined and the related mechanism analysis is used to conclude the relations between the numbers of links and joints (gear joint, revolute joint and pin-in-slot joint). The numbers of gear joints and revolute joints corresponding to various link numbers of gear trains with a pin-in-slot joint can be defined as the design specifications. The comparison between the kinematic analysis of gear trains and the astronomical theories can conclude the rates of input and output links, the design constraints and the relations of tooth counting.

2.2. Design specifications

Design specifications are the major statements of synthesized designs. Based on the information from the assigned works of historical archives, the numbers and categories of members and joints as well as the input modes should be definitely presented. Also, the category of gear train should be determined. The characteristics of topological structure are incomplete because the lost structure of gearing is without the existing design. But, the kind of generalized chains can be clearly determined.

2.3. Generalized chains

The purpose of generalization is to transform a mechanism involving various types of members and joints into a generalized chain with only generalized links and generalized joints. Following the concepts of generalization and number synthesis, the atlas of generalized chains is obtained. A generalized joint is a joint in general, and it can be any common kind of joint. A generalized link is a link with generalized joints, and it can be a link with any amount of joints. Therefore, through the process of generalization, an existing design can be generalized into the corresponding a generalized chain. By the algorithm of number synthesis, all possible generalized chains with the same numbers of links and joints as the original generalized chain can be obtained.

2.4. Specialized chains

Specialization is to assign specific categories of links and joints in the available atlas of generalized chains in accordance with the design constraints concluded from the investigation of historical archives. Through specialization and following the design constraints, a generalized chain is transformed into a feasible specialized chain. As a result, the atlas of the feasible specialized chains is obtained from the specialization for the atlas of generalized chain.

2.5. Reconstruction designs

The final step is to generate feasible reconstruction designs. At first, the atlas of feasible specialized chains is particularized into the corresponding schematic diagrams through the process of particularization. Through the teeth counting, the gears of conceptual designs can be built. Considering the geometrical constraints of surviving evidence, the feasible designs for each independent subsystem are generated. And then, whole reconstruction designs including all subsystems are obtained by considering the interference of subsystems during the process of assembly. Finally, the simulation model and physical model can be built to verify the feasibility of the reconstruction designs.

3. Design example of lost solar subsystem

In the ancient solar theory, the solar motion is clearly described through the epicycles system and the eccentric system. These two geometrical systems are equivalent to generate the solar anomaly motion, i.e., the rate of the sinusoidal function [3, 30]. As described in the work of building historical archives, the relations describing the numbers of links and joints for the lost solar subsystem are concluded by mechanism analysis of ordinary or epicyclic gear trains with a pin-in-slot joint. The numbers of mechanical elements are calculated. According to the kinematic analysis of gear trains with a pin-in-slot joint and the kinematic characteristics concluded from the ancient astronomical theories, the types of links adjacent to each other with a pin-in-slot are determined with the numbers of mechanical elements.

Without any existing evidence, it is reasonably declared that the simplest designs are the principle of our reconstruction designs for all feasible design examples. Therefore, by identifying the appropriate input conditions, three design examples of the lost solar subsystem and define the design specifications is defined as follows: (1) when the eccentric system is applied and the input power rotates counterclockwise, designs are ordinary gear trains with four links, three revolute joints, one gear joint and one pin-in-slot joint; (2) when the eccentric system is applied and the input power rotates clockwise, designs are ordinary gear trains with five links, four revolute joints, two gear joints and one pin-in-slot joint; and (3) for the epicycles system, designs of epicyclic gear train are used. The design constraint is concluded that only one of links to which the pin-in-slot joint incident can be a planet gear based on the kinematic analysis of epicyclic gear trains with a pin-in-slot joint. And thus, epicyclic gear trains with five links and seven joints (four revolute joints, two gear joints and one pin-in-slot joint) is presented [3, 30].

Following the systematic design procedure, three design examples of the lost solar subsystem respectively generate two, one and one feasible conceptual designs, as shown in Figs. 4(a)-(d). Figs. 4(a)-(c) are new designs, and Fig. 4(d) is the existing design by M. Wright [8-18]. In these mechanical drawings, the conceptual designs of the solar subsystem are assembled with the surviving structure, but the geometrical sizes and the possible pillars in the surviving gear is ignored. The inner and outer shafts respectively operate the date pointer and the Sun pointer. Figs. 4(a)-(b) show that the input link (link 2) of the reconstruction designs of the lost solar subsystem meshes with the surviving contrate gear, and the tooth and rate of the input link are variable based on the analysis of gear ratios. Figs. 4(c)-(d) show that the input link of reconstruction designs of lost solar subsystem connects to the existing gear. Therefore, the rate of input link is confirmed as once a tropic year. Except for the input link (link 2) of Fig. 4(d), other gears (links) in the reconstruction designs are hypothetical. In addition, we can calculate their appropriate teeth based on the analysis of gear ratios and the concluded kinematic characteristics of the solar theory. However, Fig. 4(d) indicates that the existing gear is

IOP Conf. Series: Materials Science and Engineering **658** (2019) 012014 doi:10.1088/1757-899X/658/1/012014

regarded as the input link (link 2) of the reconstruction designs of the lost solar subsystem to simplify the structure, once the radius of the existing gear is enough to contain the planet gears (link 5 and link 4) and the sun gear (link 1).

4. Design example of lost planetary subsystem

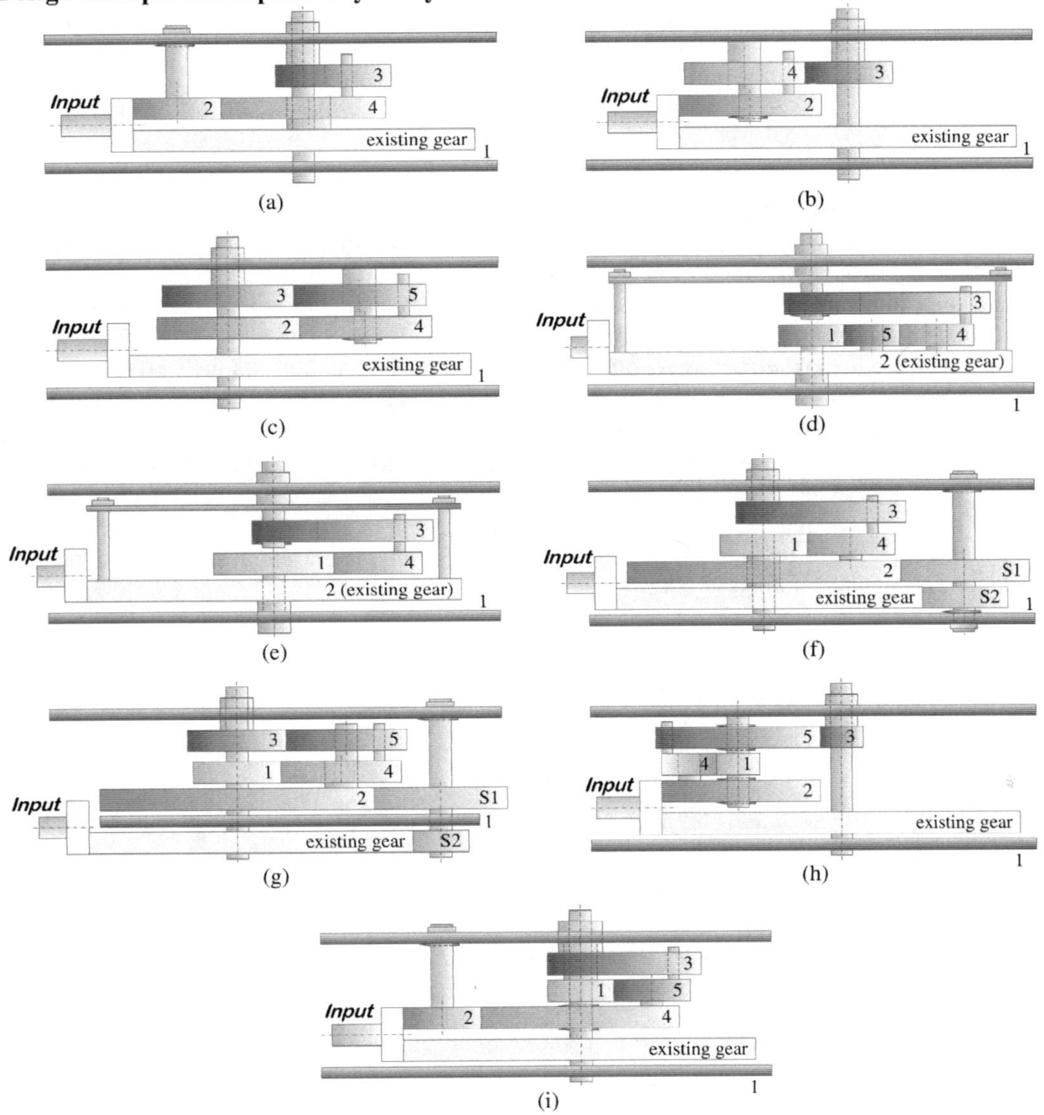

Figure 4. All feasible designs of the solar subsystem and the planetary subsystem.

Following the same way of design example of the solar subsystem, all reconstruction designs of the lost planetary subsystem can be synthesized [3, 31]. Either the interior planets (Venus and Mercury) or the superior planets (Mars, Jupiter and Saturn), designs of epicyclic gear trains with a pin-in-slot joint are used to simulate the epicycles system of ancient planetary theory. Also, the kinematic characteristics of the planetary anomaly motions are derived by the epicycles system. Mechanism and kinematic analysis of epicyclic gear trains with a pin-in-slot are contributed to conclude the design specifications, design constraints, and tooth counting relations. Furthermore, the design constraint of links, which are adjacent to each other with a pin-in-slot joint, is defined [25, 29]. According to the previous research work of historical archives, two design examples of the planetary subsystem are

IOP Conf. Series: Materials Science and Engineering **658** (2019) 012014 doi:10.1088/1757-899X/658/1/012014

concluded as follows: (1) a mechanism with four links and five joints (three revolute joints, one gear joint and one pin-in-slot joint); and (2) a mechanism with five links and seven joints (four revolute joints, two gear joints and one pin-in-slot joint).

By demonstrating design procedure and considering the input conditions of the lost planetary subsystem, these two design examples respectively generate two and three feasible reconstruction designs, as shown in Figs. 4(e)-(i). The carrier's rate of the epicyclic gear trains are related to the rate of deferent [25, 29]. Therefore, according to the periods of the epicycle and the deferent in the epicycle system, it is identified that the designs of Figs. 4(e), (g)-(i) can demonstrate the motions of the inferior planets, and the designs of Figs. 4(f), (g)-(i) can demonstrate the motions of the superior planets.

5. A reconstruction design with complete gearing

Here, the feasible designs of the solar subsystem and the planetary subsystem are integrated to generate the complete reconstruction designs of the lost structures. In order to satisfy the existing structure, some constraints about combining the individual conceptual designs are concluded based on the input conditions, the existing structure and the gear sizes (geometrical space). For the design shown in Fig. 4(a), the link 4 must rotate once a tropical year. The link 2 is an idle gear, so the teeth of link 4 must be equal to the one of existing gear (224 teeth). The size of link 4 cannot be equipped in the finite space. The designs of coaxial arbors cannot be satisfied. Therefore, only three designs of the solar subsystem, Fig. 4(b)-(d) are feasible for the integration of lost gearing. Furthermore, considering the arrangement of driving power and the type of planets, the inferior planets can have four feasible designs, Fig. 4(e)(g)-(i), and the superior planets can have two feasible designs, Fig. 4(e) and (f). As a result, 24 feasible reconstruction designs of the lost gearing integrating the solar and planetary subsystems are generated in total [3, 33].

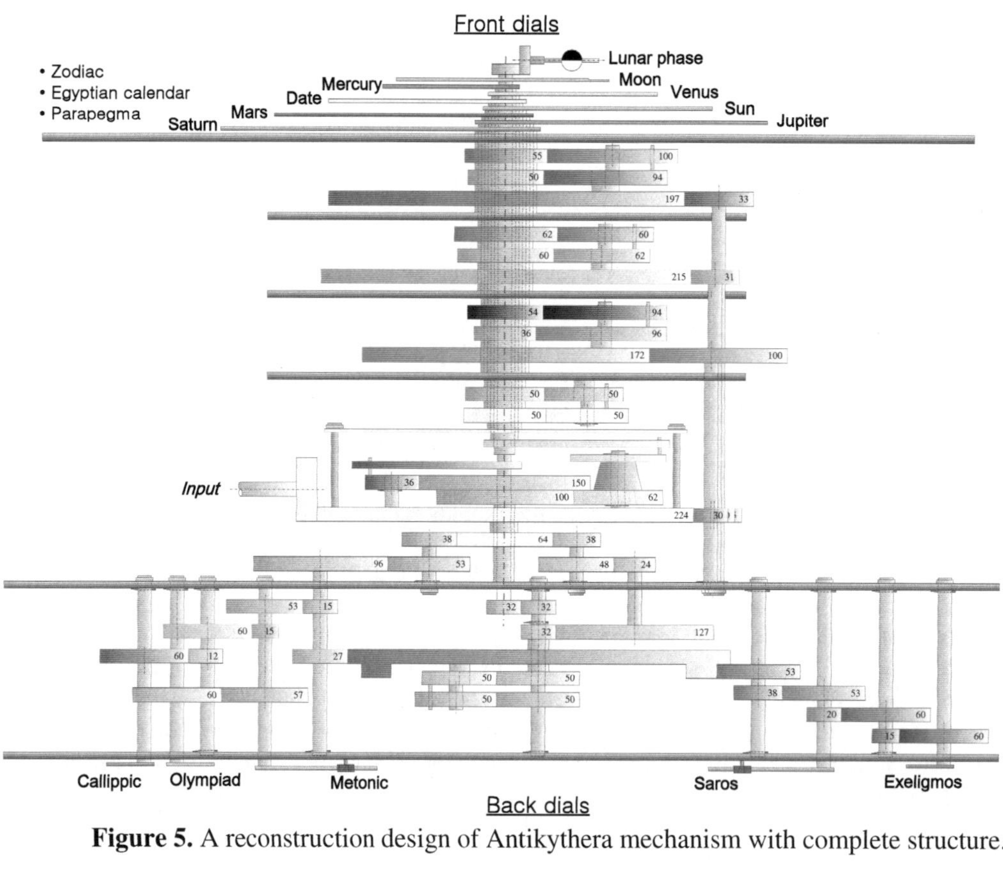

Figure 5. A reconstruction design of Antikythera mechanism with complete structure.

It is able to generate the whole reconstruction designs of Antikythera mechanism by assembling the six subsystems. Figure 5 is one of feasible reconstruction designs with complete interior structure. In its lost structures, the design is applied that the mechanisms for different celestial bodies' motions are individual. When one mechanism with a specific function is removed, the remaining mechanisms with different functions are not affected. It is hypothesized that if the future evidence indicates that the device does not demonstrate someone planet, the reconstructed structure is still workable by directly taking out the partial mechanism for the planet. Certainly, the designs of lost gearing can be simplified by utilizing the common gear (as carrier).

Finally, a simulation model and a physical model for the reconstruction design of Fig. 5 are built, as shown in Fig. 6 [33]. All gears in the lost gearing are based on the relations of tooth counting, which are concluded by kinematic analysis of gear trains in each design examples.

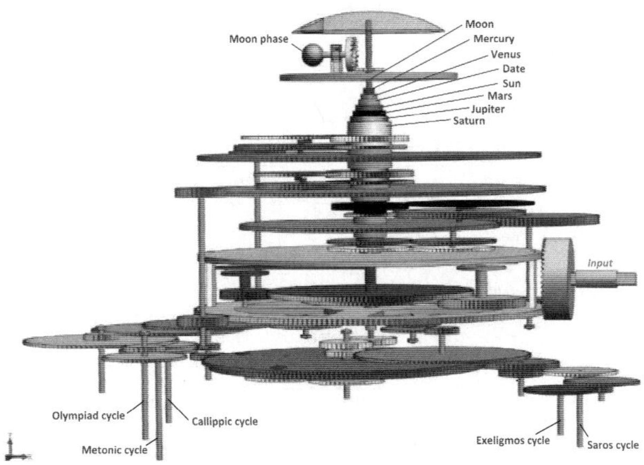

(a) simulation model

(b) physical model

Figure 6. Reconstruction models of a feasible reconstruction design.

6. Conclusions

Antikythera astronomical device is a mechanical calculator with multiple functions and complicated design, and high standard of manufacturing. The device almost completely demonstrates the developed astronomy at that time period. Among the interior structure, the lost gearing consists of the solar subsystem and the planetary subsystem. The paper reviewed the reconstruction research for the lost gearing in detail and presented some comments for the reconstruction results. The author's research presented a systematic design procedure with five steps to reconstruct the lost solar and planetary subsystems individually. During the process of integration of these two lost subsystems, geometrical constraints, interference with surviving structure, driving power arrangements, as well as types of planets were considered. The gearing for the Sun, the gearing for inferior planets and the gearing for superior planets respectively generated 3, 4 and 2 feasible reconstruction designs. And, the combination of the possible designs of lost subsystems can obtain 24 reconstruction designs of lost gearing. The arrangement of whole structure is quite flexible for the revision work, once the latest decoded information is presented in the future. In conclusion, the reconstruction research presented all feasible reconstruction designs of the lost gearing and provided new notions for decoding the complete interior structure of Antikythera astronomical device.

References

[1] Freeth T. et al. 2006 *Nature* **444** pp 581−591
[2] Freeth T. et al. 2008 *Nature* **454** pp 614−617
[3] Lin J. L. and Yan H. S. 2015 *Decoding the Mechanisms of Antikythera Astronomical Device* (Beijing: Springer)
[4] Price D. de S. 1959 *Scientific American* pp 60−67
[5] Price D. de S. 1974 *Trans Am. Phil. Soc. New Ser.* **64** pp 1−70
[6] Edmunds M. and Morgan P. 2000 *Astron. Geophys.* **41**(6) pp 10−17
[7] Edmunds M. 2014 *Contemporary Physics* **55**(4) pp 263−285
[8] Wright M. T. and Bromley A. G. 2001 *Proc. Conf. Extraordinary Machines and Structures in Antiquity* (Patras)
[9] Wright M. T. 2002 *Horological Journal* **144**(5) pp 169−173
[10] Wright M. T. 2002 *Proc. Conf. Ancient Greece and the Modern World* pp 86−97 (University of Patras)
[11] Wright M. T. 2003 *Antiquarian Horology* 1 **27** pp 270−279
[12] Wright M. T. 2004 *Bulletin of the Scientific Instrument Society* **80** pp 4−11
[13] Wright M. T. 2005 *Bulletin of the Scientific Instrument Society* **85** pp 2−7
[14] Wright M. T. 2005 *Antiquarian Horology* 2 **29** pp 51−63
[15] Wright M. T. 2005 *Bulletin of the Scientific Instrument Society* **87** pp 8−13
[16] Wright M. T. 2006 *Antiquarian Horology* **29**(3) pp 319−29
[17] Wright M. T. 2006 *Proc. Conf. Ancient Greek Technology* pp 49−60 (Athens)
[18] Wright M. T. 2007 *Interdisciphinary Science Review* **32**(1) pp 27−43
[19] Freeth T. 2002 *Mediterranean Archaeology & Archaeometry* **2** pp 21−35
[20] Freeth T. 2002 *Mediterranean Archaeology & Archaeometry* **2** pp 45−58
[21] Freeth T. 2009 *Scientific American* **301**(6) pp 79−83
[22] Freeth T. and Jones A. 2012 *The Cosmos in the Antikythera Mechanism* Institute *for the Study of the Ancient World*
[23] Koetsier T. 2008 *International Symposium on History of Machines and Mechanisms Proceedings of HMM 2008* pp 269−294 (Netherlands: Springer)
[24] Evans J., Carman C. C. and Thorndike A. S. 2010 *Journal for the History of Astronomy* **42** pp 1 −39
[25] Carman C. C., Christián C., Thorndyke A. S. and Evans J. 2012 *Journal for the History of Astronomy* **43**(1) pp 93−116
[26] Efstathiou K., Basiakoulis A., Efstathiou M., Anastatsiou M. and Seiradakis J. H. 2012 *Mechanism and Machine Theory* **52** pp 219−231

[27] Jones A. 2012 *Lecture in From Antikythera to the Square Kilometre Array: Lessons from the Ancients* pp 12－15 (Kerastari Greece)
[28] Edmunds M. G. 2014 *Contemporary Physics* **55**(4) pp 263－285
[29] Yan H. S. and Lin J. L 2011 *ASME Transactions Journal of Mechanical Design* **133**(2) p 021004
[30] Yan H. S. and Lin J. L 2013 *Mechanism and Machine Theory* **70** pp 354－371
[31] Yan H. S. and Lin J. L 2012 *ASME Transactions Journal of Mechanical Design* **134** p 011003
[32] Yan H. S. and Lin J. L 2012 *Proceedings of the Institute of Mechanical Engineers* C *Journal of Mechanical Engineering Science* **226** pp 1053－1067
[33] Lin J. L and Yan H. S. 2017 *Lecture Notes in Electrical Engineering, Mechanism and Machine Science: Proceedings of ASIAN MMS 2016 & CCMMS* **47** pp 1－14 (Springer)
[34] Yan H. S. 1992 *Mechanism and Machine Theory* **27**(3) pp 235－242
[35] Yan H. S. 1998 *Creative Design of Mechanical Devices* (Singapore: Springer-Verlag)
[36] Yan H. S. 2007 *Reconstruction Designs of Lost Ancient Chinese Machinery* (Netherlands: Springer)

IOP Conf. Series: Materials Science and Engineering **658** (2019) 012015 doi:10.1088/1757-899X/658/1/012015

Optimization of the STEP-NC compliant online toolpath generation for T-spline surfaces using convolutional neural network and random forest classifier

O Zavalnyi[1], G Zhao[1,2], Y Liu[1] and W Xiao[1,2]

[1]School of Mechanical Engineering & Automation, Beihang University, China
[2]MIIT Key Laboratory of Aeronautics Intelligent Manufacturing, Beihang University, China

E-mail: xiaowenlei@buaa.edu.cn

Abstract. Deep implementation of principals of the intelligent STEP-NC compliant manufacturing implies enabling a CNC system with integrated CAM functionality to perform all tasks autonomously. This means that most of the responsibilities of a CAM engineer need to be delegated to the CNC system. When it comes to manufacturing of freeform surfaces, the implementation of this approach becomes very challenging. One of the most important issues is the autonomous online toolpath generation. Making a choice of optimal machining strategies and other manufacturing parameters, that might be a trivial task for an expert, turns out to be not an easy problem for a machine, and often can be hardly solved using the traditional approach of explicit programming. Therefore, this paper proposes a method to optimize toolpath generation for T-spline surfaces, in particular, the process of choosing an optimal machining strategy for a given surface region using Machine Learning. The selection model has been trained to make a choice of the appropriate freeform machining strategy (two different strategies have been considered) based on the shape of an input surface.

1. Introduction

Feature oriented data representation provided by STEP-CN and the possibility of bidirectional dataflow allow to build CNC systems with integrated CAM functionality [1]. Advantages of the comprehensive integration of STEP-CN into manufacturing processes can be fully estimated while developing intelligent STEP-CNC systems. These systems are expected to perform much of their intelligent functionality (including online toolpath generation) autonomously, and the human-machine interactions have to be minimized.

Enabling such systems to machine freeform surfaces represents a significant challenge for developers. Efficient online toolpath generation for complex freeform surfaces is one of the biggest problems that need to be solved in order to build an intelligent CNC. Zhao et al. [2] presented results of the development of a STEP-compliant CNC with T-spline enabled toolpath generation capability. The reason for choosing the T-spline surface representation for integration within the STEP-CNC is its advantages over other freeform surface representations [3]. These advantages ensure more safety and flexibility during the online toolpath generation processes performed inside a CNC system, and therefore make T-spline a good choice for the development of the intelligent STEP-NC-based manufacturing.

Thus, having completed the development of a real T-spline enabled STEP-CNC system that can strategically support online toolpath generation for freeform machining of simple T-spline surfaces using different freeform strategies, this paper proposes a method to select the optimal freeform machining strategy for a surface region based on its shape using Machine Learning. This task has to be solved in order to enable a machine to make a decision about manufacturing parameters (choosing the machining strategy) based on the previous experience, imitating intellectual abilities of a human (an expert). Providing the machine with this ability allows to optimize toolpath generation, and to perform it more autonomously.

To achieve this result, the strategy selection model has been trained on a comparatively small generated training dataset (5100 pictures) of simple freeform surfaces. For each surface, an optimal machining strategy has been assigned by an expert. Two different architectures of the model have been used for training, Convolutional Neural Network (CNN) and Random Forest classifier, and their results have been compared. The developed algorithm allows to select the appropriate machining strategy (its ID) from a list for any given simple freeform surface (the input is a STL file). The input surface is being transformed into a 2D array (200x200), and the trained model can be used to perform the selection.

2. STEP-NC compliant toolpath generation for T-spline surfaces

First toolpath generation and machining results for simple T-spline surfaces were presented in Gan et al. [4]. The introduction of STEP-compliant data models for T-splines [5] made possible defining them in terms of STEP-NC and their further integration within a STEP-CNC system [2].

2.1. T-spline surface definition

T-spline surface representation, first introduced in Sederberg et al. [3], is a modification of the NURBS representation with much more topological flexibility. It has significant advantages over the NURBS such as less control points, gap-free model representation capability and localized refinement operations.

To calculate a point on a T-spline surface in three-dimensional space, the following point-based equation can be used:

$$\mathbf{S}(s,t) = \frac{\sum_{i=1}^{n} P_i B_i(s,t) w_i}{\sum_{i=1}^{n} B_i(s,t) w_i} \tag{1}$$

where Pi are the control points, w_i are weights, and $B_i(s,t)$ are the T-spline blending functions:

$$B_i(s,t) = N_i(s)\, N_i(t) \tag{2}$$

B-spline basis functions $N_i(s)$ and $N_i(t)$ are associated with the individual knot vectors in s and t directions of the parametric domain correspondingly. The components of the knot vectors s_i and t_i for each basis function are deduced from the T-mesh, which is a grid that may have T-junctions [3] (simple T-spline surfaces with T-junctions are shown in Figure 1).

2.2. STEP-NC compliant freeform machining strategies

ISO 14649-11 defines the freeform strategy as a machining strategy used for milling operations. There are four freeform strategies presented in the standard: UV strategy, Plane Cutter Contact (PCC) strategy, Plane Cutter Location (PCL) strategy, and Leading Line strategy. All these strategies have been implemented in terms of STEP-NC definitions and integrated within the developed STEP-CNC system capable to perform online toolpath generation in order to machine simple T-spline surfaces.

Unlike any other strategy, UV strategy defines the toolpaths to follow parametric lines in the local (u, v) coordinate system (Figure 1 (a)). According to the definition of the PCC strategy, the cutter contact paths can be generated by intersecting the target surface with a family of parallel planes, as shown in Figure 1(b). The only difference of the PCL strategy is that the intersections are performed with the target surface, offset by the cutter radius.

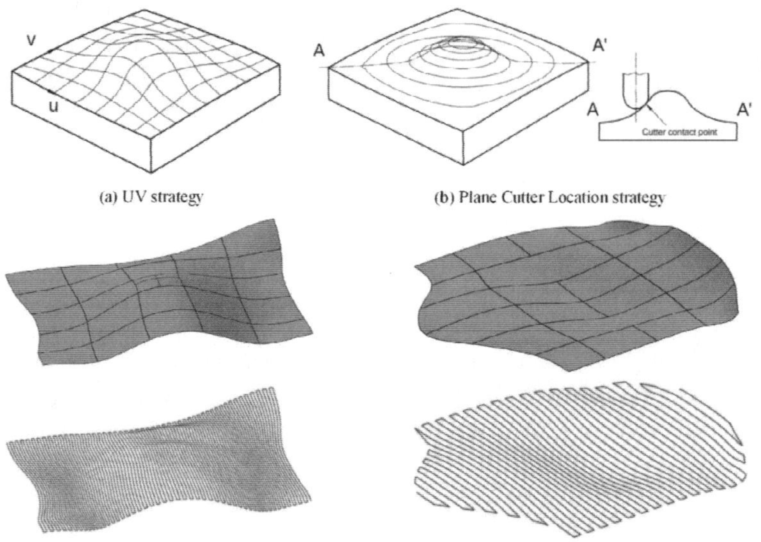

Figure 1. Toolpath generation results for simple T-spline surfaces in terms of two different freeform machining strategies defined in ISO 14649-11.

3. Scope of the investigation

Considering the complexity of the problem domain and the need to have a sufficient dataset in order train strategy selection models using Machine Learning, some limitations defining the scope of the investigation have to be imposed.

Thus, we consider only simple (in terms of the shape and topology) freeform surfaces as an input for the strategy selection algorithm. Though most of the surfaces to be machined are not simple, some existing algorithms [6, 7] allow to subdivide any complex surface into a set of feature-based regions. These region-based tool path generation methods are based on dividing the freeform surface into regions by identifying meaningful features, including surface shape and machinability. Moreover, the possibility to use local refinement operations (introduced with T-splines) makes the subdivision process even more efficient. Therefore, the problem of toolpath generation for a complex T-spline surface can be reduced to the problem of toolpath generation for the set of simple surface regions.

The number of strategies available for selection by the trained model has also been limited. In this paper, two main freeform machining strategies presented in the standard have been considered, UV strategy and PCC strategy. These two strategies have one significant difference which makes it very important to choose the one strategy and not the other in each specific case. Under the UV strategy, toolpath lines are defined by iso-parametric lines. Therefore, despite that the distance between two neighboring toolpath lines in the parametric domain is equal, this distance may vary considerably in the 3D space depending on the surface shape and topology. Considering the need to maintain the constant scallop-height between consecutive trajectories on the manufacturing part, the toolpath length (and machining time) may increase. Whereas, toolpaths generated using the PCC strategy do not have this problem because cutter contact paths are defined as the intersection of the target surfaces with parallel planes in 3D space. However, toolpath generation in terms of the UV strategy is simpler and computationally cheaper than it is for the PPC strategy. Thus, the choice of the optimal machining strategy for each specific case has to be made based on the surface region shape.

4. Brief introduction to machine learning

Machine Learning is the field of study that gives computers the ability to learn without being explicitly programmed [8]. It allows a computer program to learn from experience and improve its performance with respect to some task. Recently, Machine Learning has been experiencing a never seen increase in the number of its applications in different fields, including manufacturing [9]. Deep learning is part of

IOP Conf. Series: Materials Science and Engineering **658** (2019) 012015 doi:10.1088/1757-899X/658/1/012015

a broader family of machine learning methods, and having as its core different architectures of artificial neural networks (ANNs), it allows to solve very complex Machine Learning tasks.

The problem of the machining strategy selection is a classification problem. After training the model on a set of the labelled training data, this model can be used to define the class of any new sample. Many different model architectures could be used to solve this problem, and in this paper Convolutional Neural Network (CNN) and Random Forest classifier have been considered. CNNs have provided state-of the-art results in many computer vision problems [10], but they usually require a large number of training samples. Random Forest classifiers can also directly classify instances into multiple classes [8], and it is often possible to achieve good accuracy results even on a smaller number of training samples.

Compared to the traditional Artificial Neural Networks (ANNs), CNNs are primarily designed and used in the field of pattern recognition within images [11]. Therefore, some image-specific features, which make the network more suited for image-focused tasks, can be encoded into the architecture, and the parameters required to set up the model can be further reduced. The layers within the CNN are comprised of neurons organized into three dimensions, the spatial dimensionality of the input (height and the width) and the depth (the third dimension of an activation volume). CNNs usually contain three types of layers: convolutional layers, pooling layers and fully-connected layers [12]. The convolutional layers are used to extract features from the input images, then the dimensionality can be reduced by the pooling layers, and the fully connected layers act as classifiers. If a sufficient training dataset is available, CNNs almost always will provide the best results in pattern recognition and image classification tasks.

Random Forests classifiers are classifiers based on decision-tree that use ensemble algorithms (algorithms which combine more than one algorithms of same or different kind for classifying objects) [13]. They operate by constructing a number of decision trees at training time and outputting the class that is the mode of the classes of the individual trees. The training process is generally carried out via the bagging method [8]. Random Forests are well suited for both classification and regression tasks, and therefore may be used in various machine learning applications.

5. Preparation of the dataset for training

Proposed strategy selection algorithm takes as an input a surface given in STL format (often used format in CAM applications). The integrated within the developed STEP-CNC system T-spline kernel supports STL format for the data exchange, therefore the algorithm can be easily implemented inside the system architecture in the future. The geometrical information contained in the STL files cannot be directly used in the training processes, because different surfaces may be represented by the different number of triangles. The input for the most of models has to be a fixed-size sequence of features (a fixed number of input neurons for CCNs). Therefore, in order to train strategy selection models, the original surface has to be represented as an array of fixed size. The best way to do this is to transform the surface into a voxelized 3D model located within a grid box of fixed size. Then the dimensionality of this box can be reduced, and the model can be represented as a 2D matrix. This approach is possible due to the assumption that the input is a simple surface, and therefore it can be easily mapped on the XY-plane of the local coordinate system. By subdividing a complex surface into a set of simple surface regions, we can fulfil these requirements in the most cases. Otherwise, the 3D matrix representation can be used, and 3D convolutional neural networks have to be employed correspondently.

As an example, the transformation of a grid box (15x15x15) with one single voxel into a 2D matrix (15x15) is presented in Figure 2. X and Y-coordinates of the voxel in the box define the row and the column numbers in the matrix (12 and 8 correspondently). Z-coordinate of the voxel, 11, defines the normalized value, Zn, of the corresponding element of the matrix ($0 \leq Zn \leq 1$): $Zn = (1/15) \times 11 = 0.73$. The rest of the elements of the matrix have zero values.

In order to be used for training, each machining strategy is assigned a unique ID number (0 – for UV strategy and 1 – for PCC strategy). These ID numbers can be used to label training samples.

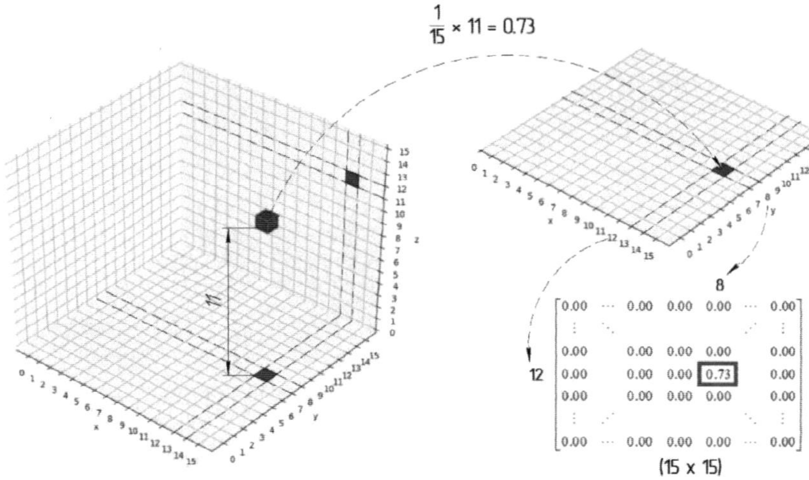

Figure 2. Transformation of a grid box with one voxel into a 2D matrix.

First, 340 unique simple T-spline surfaces of different size and shape have been created and saved in the STL format (sample surfaces are shown in Figure 3 (right)).

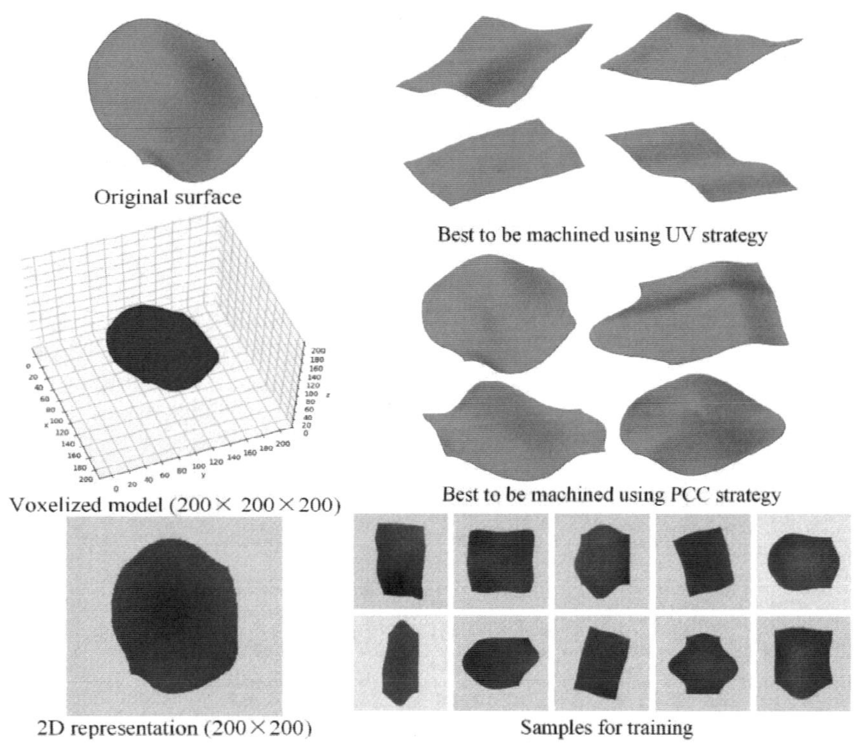

Figure 3. Transformation of the surface representation (left). Examples of different surfaces and prepared training samples (right).

For each surface, the optimal machining strategy has been assigned by an expert, and each sample was labelled with the corresponding strategy ID number. Then, the surfaces have to be transformed into the 2D matrix representations, as described in this section (see Figure 3 (left)). A grid box (200x200x200) with the cell size in all three dimensions equal to 0.5 mm is created, and the surface, being voxelized, is centered in the box. All surfaces must be scaled to fit into the box. At the same time, augmentation (by rotating the surface about the X, Y, and Z –axis by a small random angles) is performed to expand the number of training samples. To manipulate the surface transformations, open source package "trimesh" written in Python has been used. Finally, each sample is represented as a 2D matrix (200x200), which can be viewed as a fixed-size image (see Figure 3). Thus, the final training dataset after augmentation contains 5100 images of the size (200x200) and 5100 labels which define the corresponding strategy IDs.

6. Training of the strategy selection models and the test results

To perform the training, Tensorflow library has been used. Different architectures and combinations of hyper-parameters have been tested to get good results.

The CNN-based training model is a 5-layer neural network with 3 convolution layers, 200x200x1 input layer and the output layer of size 2 (two classes of the machining strategy). The model has the following architecture: one channel input layer, convolution layer of output depth 4, convolution layer of output depth 8, convolution layer of output depth 16, fully connected layer, output layer (one-hot encoded labels). The (2 × 2) kernel has been. The learning rate was set to 0.003, the activation function was a ReLU, and the Adam optimizer has been used during the training process.

To train the model based on the Random Forest classifier, the Tensorflow contributed library has been used. Its model has the following structure: the input is a vector of 40000 features (from the 200x200 input matrix), the output has 2 classes (2 strategies) and number of trees is 35 (with limited to maximum 400 nodes). However, different numbers of trees and nodes may give the same results.

To estimate the accuracy of the trained models, several new sets of test samples have been created and labelled by an expert. The models demonstrated different results for different sets of test data, as shown in Table 1:

The model based on the Random Forest classifier demonstrated slightly better results. Moreover, as the training dataset is comparatively small, the CNN-based model is very prone to overfitting and therefore requires more tuning manipulations.

Table 1. Test results for two different models.

Model type	Accuracy, %
CNN (5 layers)	85 – 94,8
Random Forest classifier	87,5 – 95.8

Sometimes, a surface may have the shape, based on which it is hard to conclude that the one machining strategy is an exclusive and the only possible choice, and the other is totally wrong (multiple choice is possible). In this case, the trained model might make a "wrong decision", but it still can be an acceptable choice in terms of machining results. Considering this fact, it is concluded that the achieved accuracy results are satisfactory and potentially may be used for the online toolpath generation applications. However, to achieve better and more stable results, a larger amount of training data has to be used.

7. Conclusions

The results of this research demonstrate the real possibility to use Machine Learning for the machining strategy selection. The trained model, after further optimization (possibly, retraining on larger datasets) can be integrated inside the STEP-CNC system and be used during the online toolpath generation. As the future work, authors consider adding more machining strategies available for the selection. Training modes for the automated selection of cutting tools and some other manufacturing parameters might be also considered in the future. Presented in this paper results are very important because they show that Machine Learning can potentially help to transfer the responsibilities to choose any manufacturing parameters from a human to a machine, as long as a sufficient training datasets are available.

Acknowledgements
This work was supported by Special Program of Ministry of Industry and Information Technology, China.

References

[1] Newman S T, Allen R D and Rosso Jr R S U 2003 *Int. J. of Computer Integrated Manufacturing* **16** pp 590－597
[2] Zhao G, Liu Y, Xiao W, Zavalnyi O and Zheng L 2018 *Int J Adv Manuf Technol* **94** pp 1799－1810
[3] Sederberg T W, Zheng J, Bakenov A and Nasri A 2003 *ACM Transactions on Graphics* **22** pp 477－484
[4] Gan W F, Fu J Z, Shen H Y, Chen Z Y and Lin Z W 2014 *Computer-Aided Design* **52** pp 51－63
[5] Xiao W, Liu Y, Li R, Wang W, Zheng J and Zhao G 2016 *Computer-Aided Design* **79** pp 36－47
[6] Chen Z C, Dong Z and Vickers G W 2003 *Computers in Industry* **50** pp 319－331
[7] Elber G 1995 *Computer-Aided Design* **27** pp 465－470
[8] Géron A 2017 *Hands-On Machine Learning with Scikit-Learn and TensorFlow* O'Reilly Media Inc. p 541
[9] Wuest T, Weimer D, Irgens C and Thoben K D 2016 *Production & Manufacturing Research* **4** pp 23－45
[10] Keshari R, Vatsa M, Singh R and Noore A 2018 Learning Structure and Strength of CNN Filters for Small Sample Size Training *CoRR* abs/1803.11405
[11] O'Shea K and Nash R 2015 An Introduction to Convolutional Neural Networks *CoRR* abs/1511.08458
[12] Canziani A, Paszke A and Culurciello E 2016 An Analysis of Deep Neural Network Models for Practical Applications *CoRR* abs/1605.07678
[13] Ho T K 1995 Random Decision Forests *in Proceedings of the 3rd International Conference on Document Analysis and Recognition* **1** pp 278－282

IOP Conf. Series: Materials Science and Engineering **658** (2019) 012016 doi:10.1088/1757-899X/658/1/012016

NOx emission from a sulfur recovery unit thermal reactor with rounded corners or a vector wall

Chun Lang Yeh

Department of Aeronautical Engineering, National Formosa University, Taiwan

E-mail: clyeh@nfu.edu.tw

Abstract. A sulfur recovery unit (SRU) thermal reactor is important equipment in a sulfur plant. Because its operating temperature is very high, the nitrogen oxides (NOx) produced by a SRU thermal reactor is one of the possible sources of air pollution. In this paper, NOx emission from a practical SRU thermal reactor is analyzed. Both the prototype thermal reactor and its modifications, including modifying the reactor geometry and using a vector wall are compared. Simulation results show that the SRU thermal reactor with a radius of curvature 1m at the zone 1 corner has the lowest NO emission. Compared with using a choke ring, the NO concentration at the reactor exit is higher using a vector wall.

1. Introduction

A SRU thermal reactor is the most important equipment in a sulfur plant. It converts the NH_3, H_2S and hydrocarbons in the reactants into sulfur. Most of the sulfur elements are recovered from the SRU thermal reactor. The first section of a SRU that uses the Claus process is composed of a burner, a thermal reactor and a waste heat exchanger. The configuration and dimensions of the first section of a SRU for a typical petroleum refinery are shown in Figure 1.

SRU thermal reactors are negatively affected by high temperature operations because high temperature can damage the refractory and the heat exchanger tubes. Our experience of operating a practical SRU thermal reactor in Taiwan shows that the refractories at the zone 1 corner and the choke ring are the parts of the thermal reactor that experience the greatest deterioration. The zone 1 corner has a suddenly expanded geometry and a recirculation zone forms behind it. The temperature at the zone 1 corner can exceed the maximum service temperature of the refractory and cause collapse or deformation. The choke ring is subjected to a bending moment from the rapid combustion gas stream and can collapse or deform.

NOx is a pollutant. It is one of the major sources of air pollution. NOx is a precursor for photochemical smog, contributes to acid rain, and causes ozone depletion. In addition, NOx has been recognized as one of the major causes of excessive peroxide concentration in the atmosphere. Thus, there is a common consensus that NOx control has become a worldwide problem. The formation of NOx can be attributed to four distinct chemical kinetic processes: thermal NOx, prompt NOx, fuel NOx, and intermediate N_2O. In general, the contribution from prompt NOx and intermediate N_2O is minor. Fuel NOx is produced by oxidation of nitrogen atoms contained in nitrogen-bearing fuels. Thermal NOx is formed by the oxidation of atmospheric nitrogen molecules present in the combustion air and strongly depends on the temperature. Considering the fuel composition for this study shown in Table 1, thermal NOx is the major source of NOx in this study. NOx emission consists of mostly nitric oxide (NO) (around 95%), and to a lesser degree nitrogen dioxide (NO_2) and nitrous oxide (N_2O) (around 5%). Therefore, the NO species is the major concern in this study.

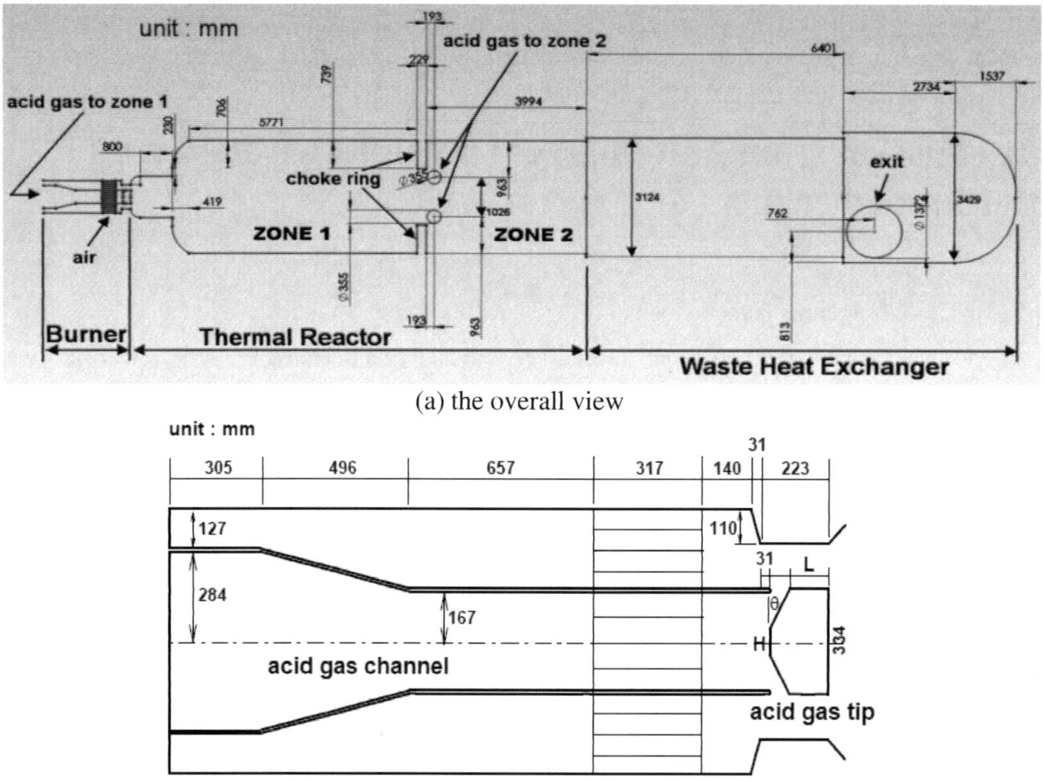

(a) the overall view

(b) enlarged view for the burner section

Figure 1. The configuration and dimensions of the first section of the SRU for a typical petroleum refinery

The flow field in a SRU thermal reactor involves many complicated mechanisms, such as turbulent mixing, convection and radiation heat transfer modes, combustion, as well as NOx and SOx formations. These mechanisms are all highly complex and may interact with each other, which makes the analysis much more difficult. Hassan et al. [1] simulated the combustion process occurring in the combustion chamber of some domestic boilers. A two-dimensional CFD model is built to simulate the combustion chamber domain, and the partially premixed combustion model with a postprocessor for NOx calculations is used to simulate the combustion process inside the combustion chamber. Franco and Diaz [2] described the state of the art in the field of ''clean coal technologies'' showing the perspectives of improvement and the critical elements. They reviewed and analyzed the emission control of NOx and SOx, and particle matter as well as advanced coal conversion pathways such as ultra-supercritical (USC), pressurized fluidized bed combustion (PFBC), and integrated gasification combined cycle (IGCC). Normann et al. [3] reviewed available techniques for controlling both the emission of NOx to the atmosphere and the content of NOx in the captured carbon dioxide. Shin et al. [4] developed a comprehensive computer program to evaluate the efficiency of a selective non-catalytic reduction (SNCR) system for a boiler. The standard k-ε turbulence model and eddy breakup model were incorporated to analyze the Reynolds stresses and the turbulent reaction of major fuel species, respectively. Javed et al. [5] presented a review of NOx removal techniques with particular reference to SNCR technology. A review of various features related to selective non-catalytic gas phase injection of ammonia and ammonium salts (as reducing agent) was presented. CFD modeling was also applied to SNCR.

In this paper, NOx emissions from a practical SRU thermal reactor are analyzed. The purpose of this study is to compare the NOx emissions from the SRU thermal reactors (both prototype and its modifications). Practical operating conditions from a petrochemical corporation in Taiwan were used as the design conditions for the discussion.

2. Numerical methods and physical models

In this study, the FLUENT commercial code is used to simulate the reacting and fluid flow in a SRU thermal reactor. The SIMPLE algorithm by Patankar[6] is used to solve the governing equations. The discretizations of convection terms and diffusion terms are respectively performed using the power-law scheme and the central difference scheme. In terms of physical models, considering the accuracy and stability of the models and the evaluations of other researchers, the standard k-ε Model [7], the P-1 radiation model [8] and the non-premixed combustion model with β–type probability density function [9] are respectively used for the turbulence, radiation and combustion simulations. The standard wall functions [10] are used to resolve the flow quantities (velocity, the temperature, and the turbulence quantities) at the near-wall regions. The detailed governing equations and convergence criterion were described in previous studies by the author [11, 12].

In this study, the numerical model of a SRU thermal reactor is constructed using an unstructured grid. Figure 2 shows the numerical model of the prototype SRU thermal reactor. In Figure 2, the heat exchanger section consists of 19 cooling tubes, which have a diameter of 0.5m, as shown schematically in Figure 3. The heat absorption rate for each heat exchanger tube is 40,000 W/m^2 and the other walls are adiabatic. No slip condition is applied on any of the solid walls. The exit of the heat exchanger section is connected to other equipment at 300 K and 1 atm by a pipe that is 1.372m in diameter and 11.5m in length.

Figure 2. Numerical model for the SRU thermal reactor investigated

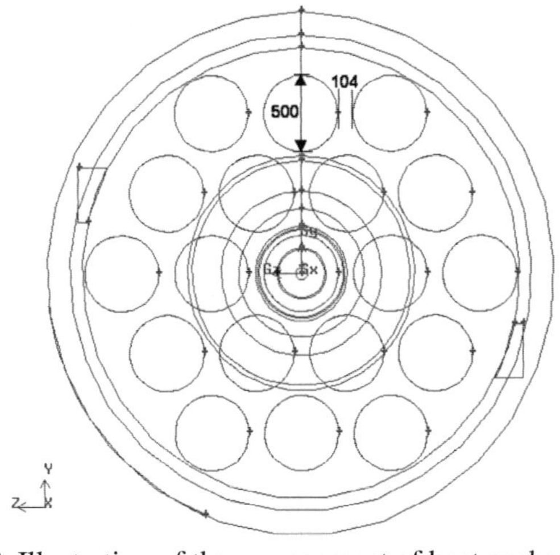

Figure 3. Illustration of the arrangement of heat exchanger tubes

In this study, two types of oxygen supplies are investigated: an oxygen-normal supply and an oxygen-rich supply. An oxygen-rich supply increases sulfur recovery. The design conditions (including the species compositions, the temperature, the pressure and the velocity) at the acid gas inlet holes of zone 1 and zone 2 and at the air inlet hole are listed in Table 1. These conditions are practical operating conditions that are used by a petrochemical corporation in Taiwan. The turbulence kinetic energy is 10% of the inlet mean flow kinetic energy and the turbulence dissipation rate is computed using Eq.(1).

$$\varepsilon = C_\mu^{3/4} \frac{k^{3/2}}{l} \tag{1}$$

where l=0.07L and L is the hydraulic diameter.

Table 1. The design conditions at the acid gas inlet holes and the air inlet hole

Location	Oxygen-Normal Supply			Oxygen-Rich Supply		
	Acid Gas to Zone 1	Acid Gas to Zone 2	Air Inlet	Acid Gas to Zone 1	Acid Gas to Zone 2	Air Inlet
Species					x (%)	
O_2	0	0	19.87	0	0	23.85
N_2	0	0	74.98	0	0	71.26
H_2O	7.83	4.12	5.15	4.12	27.97	4.89
CO_2	1.27	1.5	0	1.48	0	0
H_2S	82.06	89.88	0	89.9	39.61	0
CH_4	2.28	2.7	0	2.7	0	0
C_2H_6	1.52	1.8	0	1.8	0	0
NH_3	5.04	0	0	0	32.42	0
T (K)	319.92	316.15	403.15	313.15	316.15	397.15
P (N/m^2)	76,920	75,068	74,382	75,068	75,068	89,572
V (m/s)	11.62	2.08	12.4 (Radial)	11.46	1.88	10.8 (Radial)
title			34.1 (Tangential)			29.8 (Tangential)

The grid independence test and validation of the numerical methods have been performed in the author's previous studies [11, 12].

3. Results and discussion

3.1. Effect of reactor geometry

Table 2 shows the exit NO concentrations for SRU thermal reactors with different radii of curvature at the zone 1 corner. Five different radii of curvature at the zone 1 corner, including 0m (i.e. without streamlining), 0.5m, 1m, 1.5m and 2m, are calculated to investigate the geometric effects of the zone 1 corner. Figure 4 shows the numerical models of a SRU thermal reactor with different radii of curvature at the zone 1 corner. From the author's previous study [13], it was found that the lowest peak temperature is obtained with a radius of curvature 1m at the zone 1 corner. Figure 5 shows the temperature profiles for the SRU thermal reactors using a radius of curvature 1m at the zone 1 corner [13]. With a streamlined zone 1 corner, the average temperature is increased due to the compression effect caused by the decreased volume in zone 1. However, the corner recirculation zone in zone 1 becomes smaller due to the streamlining effect, which yields a reduction in temperature. These two effects (compression effect and a smaller corner recirculation zone) lead to an optimal radius of curvature at zone 1 corner. With larger radii of curvature at zone 1 corner, the compression effect overwhelms the effect of a smaller corner recirculation zone and therefore the peak temperature is higher.

From Table 2, it is observed that the SRU thermal reactor with a radius of curvature 1m at the zone 1 corner has the lowest NO emission because of its lowest peak temperature.

Table 2. Simulation results for SRU thermal reactors with different radii of curvature at the zone 1 corner

(a) oxygen-normal supply

radius of curvature (m)	peak temperature (K)	average temperature (K)	Exit sulfur concentration (mole fraction)	Exit NO concentration (ppm)
0	1902	1500	0.0791	1.18
0.5	1895	1502	0.0792	0.99
1.0	1890	1505	0.0793	0.93
1.5	1913	1508	0.0795	2.45
2.0	1942	1515	0.0798	3.16

(b) oxygen-rich supply

radius of curvature (m)	peak temperature (K)	average temperature (K)	Exit sulfur concentration (mole fraction)	Exit NO concentration (ppm)
0	2103	1595	0.0902	14.7
0.5	2092	1601	0.0911	13.7
1.0	2071	1609	0.0915	13.5
1.5	2097	1610	0.0915	14.9
2.0	2111	1611	0.0915	16.1

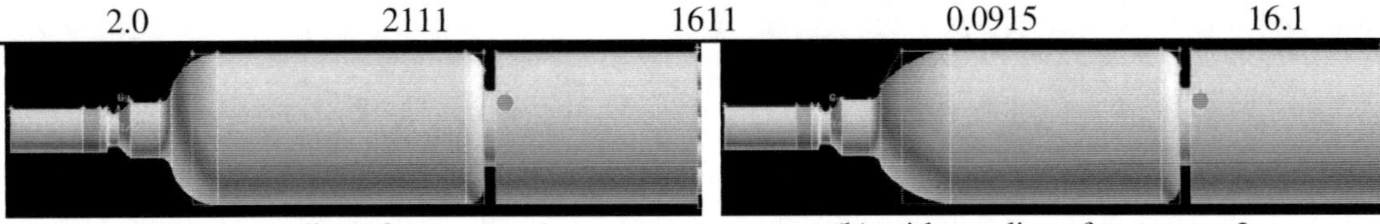

(a) with a radius of curvature 1m (b) with a radius of curvature 2m

Figure 4. Numerical models of a SRU thermal reactor with different radii of curvature at the zone 1 corner

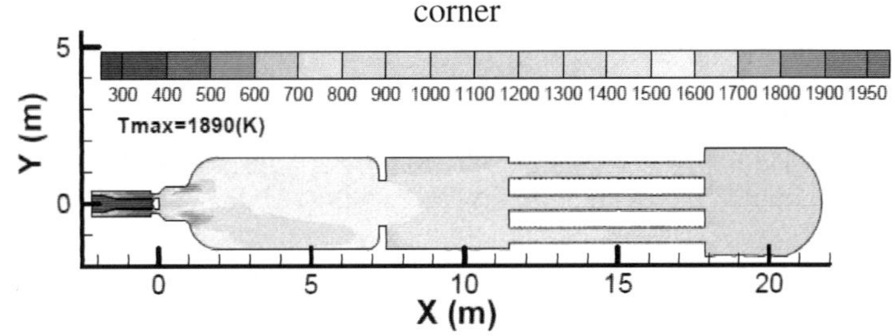

(a) oxygen-normal supply

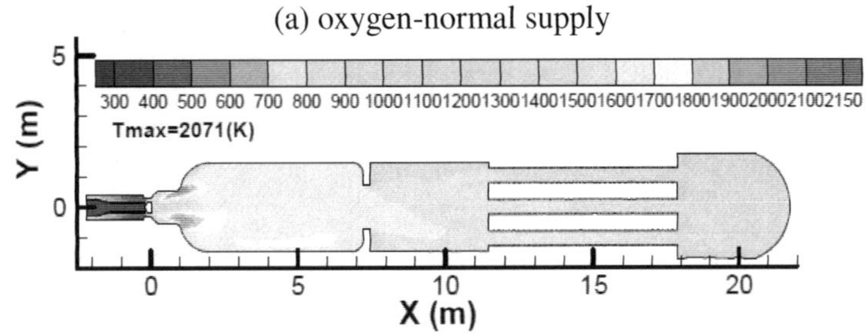

(b) oxygen-rich supply

Figure 5. Temperature profiles for the SRU thermal reactors using a radius of curvature 1m at the zone 1 corner

3.2. Effect of a vector wall

The specific arrangement of the holes of a vector wall results in spiral motion behind the vector wall and thereby increases the residence time. In addition, a vector wall enhances mixing and temperature uniformity.

(a) using a choke ring

(b) using a 36-hole vector wall

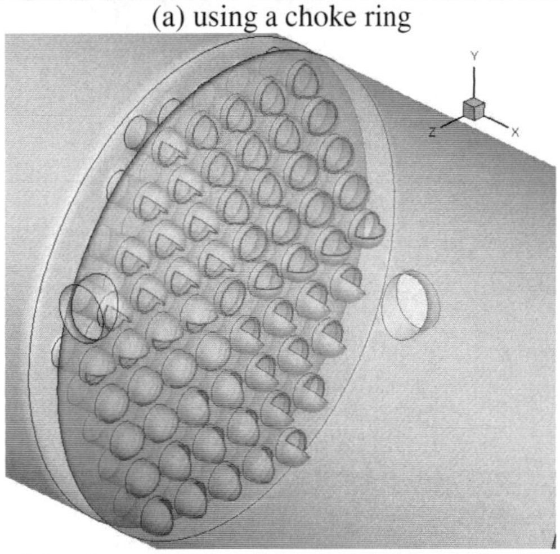

(c) enlarged view of the 36-hole vector wall

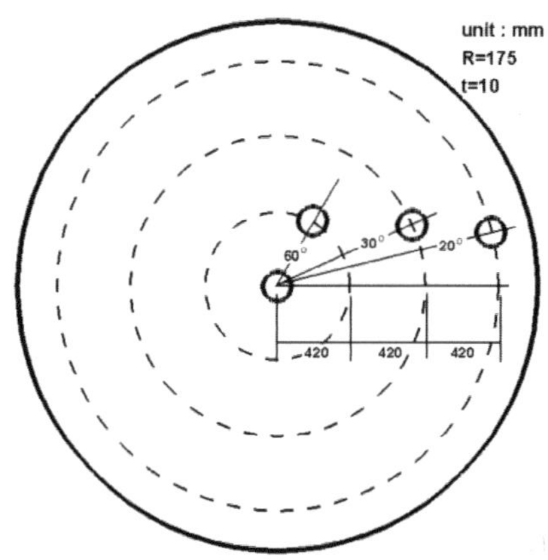

(d) arrangement of a 36-hole vector wall

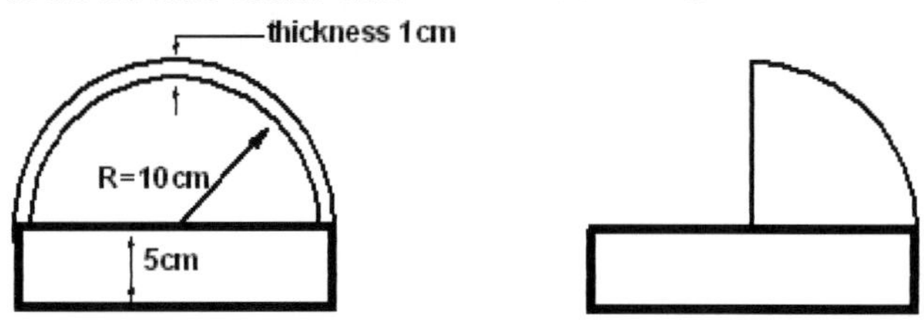

(e) configuration and dimension of a vector wall hole

Figure 6. Numerical models of a SRU thermal reactor with a choke ring or a 36-hole vector wall

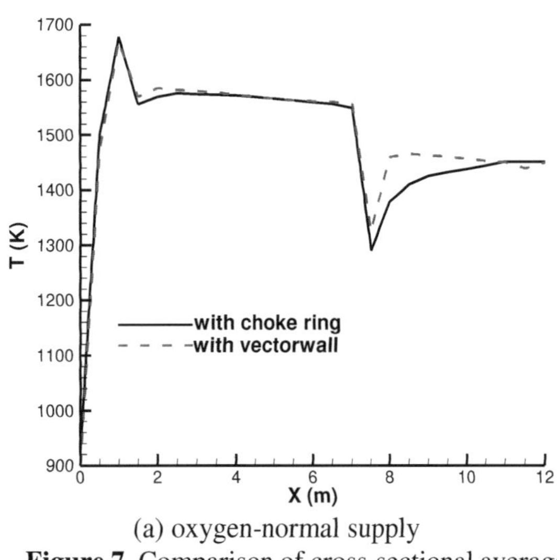

(a) oxygen-normal supply

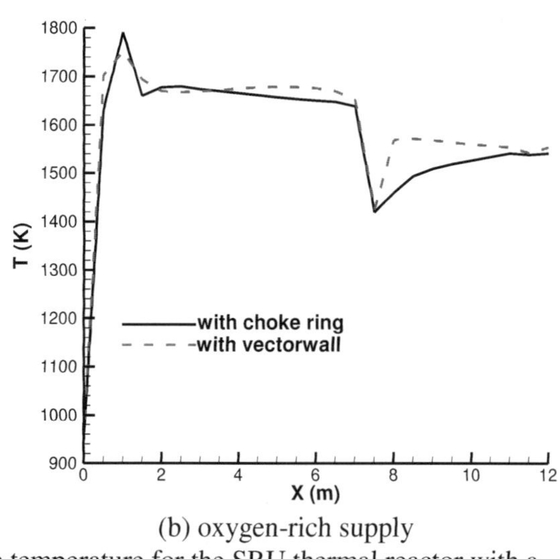

(b) oxygen-rich supply

Figure 7. Comparison of cross-sectional average temperature for the SRU thermal reactor with a choke ring or a vector wall

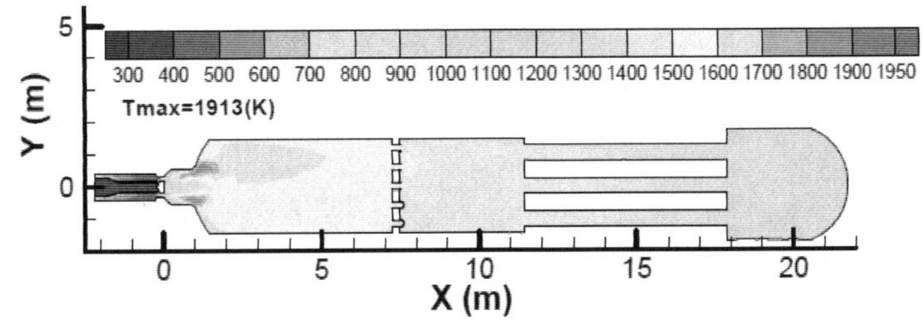

(a) oxygen-normal supply

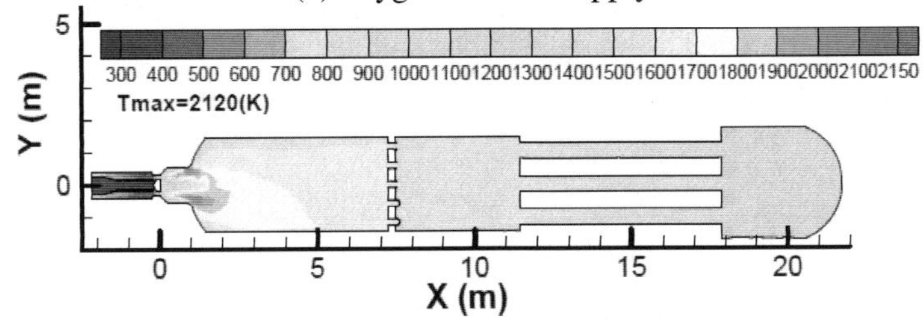

(b) oxygen-rich supply

Figure 8. Temperature profiles for the SRU thermal reactors using a vector wall

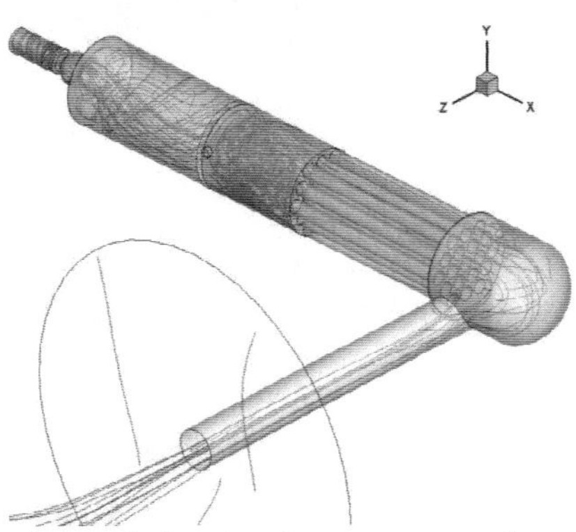

(a) using a choke ring (oxygen-normal supply)

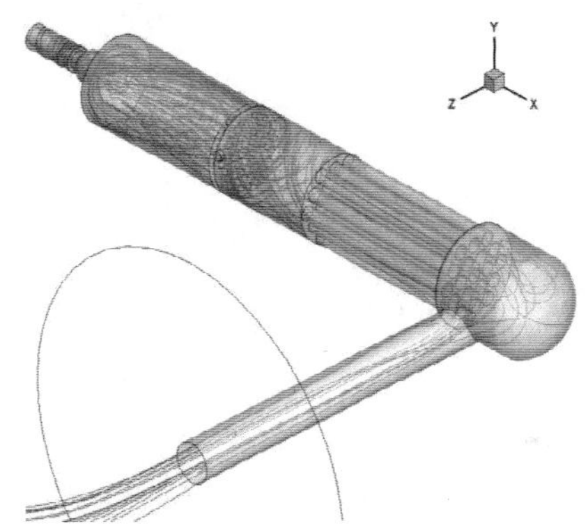

(b) using a vector wall (oxygen-normal supply)

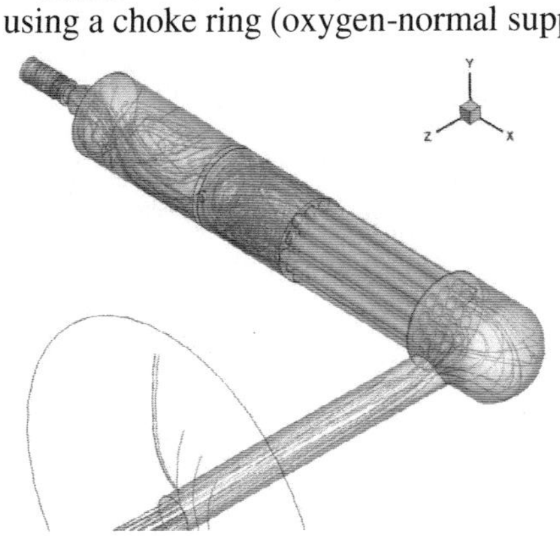

(c) using a choke ring (oxygen-rich supply)

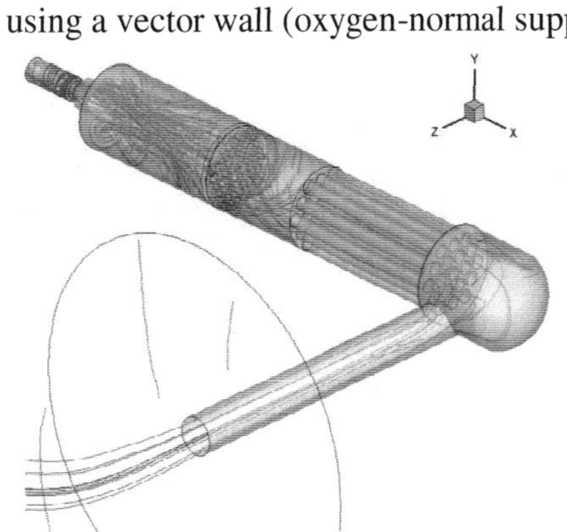

(d) using a vector wall (oxygen-rich supply)

Figure 9. Stream traces for the SRU thermal reactors using a choke ring or a vector wall

Figure 6 shows the numerical models of a SRU thermal reactor with a choke ring or a 36-hole vector wall. The choke ring or vector wall is located at 6m away from the zone 1 corner. From the author's previous study [13] (Figure 7), it was found that with a vector wall, the average temperature in zone 2 is increased due to better mixing. However, the larger surface area of a vector wall results in a larger blockage effect and hence the average temperature in zone 1 is also increased. Figure 8 shows the temperature profiles for the SRU thermal reactors using a vector wall [13]. It is observed that the average temperature across a vector wall becomes more uniform due to better mixing. Figure 9 shows the stream traces for the SRU thermal reactors using a choke ring or a vector wall [13]. The spiral motion behind the vector wall is clearly observed.

Table 3 shows the sulfur and the NO concentration at exit. In a practical SRU thermal reactor, the refractory may be ruptured due to high temperature, for example, near the zone 1 corner. It can be seen from Table 3 that the sulfur and NO concentration at exit are higher using a vector wall. The exit sulfur concentrations are increased by 0.6% and 2.5% for the oxygen-normal operation and the oxygen-rich operation, respectively, while the exit NO concentrations are increased by 14.4% and 16.3% for the oxygen-normal operation and the oxygen-rich operation, respectively.

Table 3. Sulfur and NO concentration at the exit for SRU thermal reactors with a choke ring or a vector wall

(a) oxygen-normal supply

	Exit sulfur concentration (mole fraction)	Exit NO concentration (ppm)
choke ring	0.0791	1.18
vector wall	0.0796	1.35

(b) oxygen-rich supply

	Exit sulfur concentration (mole fraction)	Exit NO concentration (ppm)
choke ring	0.0902	14.7
vector wall	0.0925	17.1

4. Conclusion

In this paper, NOx emissions from the SRU thermal reactors are analyzed. Both the prototype thermal reactor and its modifications, including modifying reactor geometry and using a vector wall, are compared. Practical operating conditions from a petrochemical corporation in Taiwan were used as the design conditions for the discussion. From the simulation results, it is found that the SRU thermal reactor with a radius of curvature 1m at the zone 1 corner has the lowest NO emission. The NO concentration at exit is higher using a vector wall in comparison with using a choke ring. The result of this study is helpful to the design of a low-nox SRU thermal reactor.

Acknowledgements

The author gratefully acknowledges the grant support from the Ministry of Science and Technology, Taiwan, R.O.C., under contract MOST 107-2221-E-150-012. The author is also grateful to the Formosa Petrochemical Corporation in Taiwan for providing valuable data and constructive suggestions to this research during the execution of the industry-university cooperative research project under the contract 104AF-86.

References

[1] Hassan G, Pourkashanian M, Ingham D, Ma L and Taylor S 2009 *J. Therm. Sci. Eng. Appl.* **1** 011007

[2] Franco A and Diaz A R 2009 *Energy* **34**(3) pp 348−354

[3] Normann F, Anderson K, Leckner B and Johnson F 2009 *Prog. Energy Combust. Sci.* **35**(5) pp 385−397

[4] Shin M S, Kim H S and Jang D S 2007 *Appl. Therm. Eng.* **27**(17-18) pp 2850−2857

[5] Javed M T, Irfana N and Gibbs B M *2007 J. Environ. Manage.* **83**(3) pp 251−489

[6] Patankar S V 1980 *Numerical heat transfer and fluid flows* (New York: McGraw-Hill)

[7] Launder B E and Spalding D B 1972 *Lectures in mathematical models of turbulence* (London England: Academic press)

[8] Siegel R and Howell J R 1992 *Thermal radiation heat transfer* (Washington DC: Hemisphere publishing corporation)
[9] Sivathanu Y R and Faeth G M 1990 *Combust. Flame* **82**(2) pp 211−230
[10] Launder B E and Spalding D B 1974 *Comput. Method Appl. Mech. Eng.* **3**(2) pp 269−289
[11] Yeh C L 2013 *Int. J. Heat Mass Tran.* **59** pp 172−190
[12] Yeh C L 2016 *Appl. Sci.* **6**(11) 331 doi:10.3390/app6110331
[13] Yeh C L 2016 *T. Can. Soc. Mech. Eng.* **40**(5) pp 811−820
[14] Yeh C L 2016 *T. Can. Soc. Mech. Eng.* **40**(4) pp 511−520
[15] Yeh C L 2015 *Int. J. Mech. Eng. Rob. Res.* **4**(3) pp 273−277

IOP Conf. Series: Materials Science and Engineering **658** (2019) 012017 doi:10.1088/1757-899X/658/1/012017

Relationship between air pollution and outpatient visits for nonspecific conjunctivitis in Taiwan: an application of Bayesian regression model

Yi Horng Lai

Oriental Institute of Technology, Taiwan

E-mail: FL006@mail.oit.edu.tw

Abstract. As a direct exposed part of the human body, the eyes are susceptible to external stimuli of air pollution. The adverse effects of air pollution on the human eye are mainly stimulation and inflammation, and conjunctivitis is a major problem. This study investigated the effects of air pollution on non-specific conjunctivitis based on data from a Taiwanese outpatient clinic. The study focused on the importance of air pollution to eye health. Data for ophthalmology outpatient visits were obtained from the National Health Insurance Research Database of Taiwan. Ambient air monitoring data were obtained from the Taiwan Environmental Protection Administration air pollution monitoring stations. The study included five regulated air pollutants. This study used Bayesian linear regression to investigate the association between daily air pollutant concentrations and the number of outpatient visits for nonspecific conjunctivitis. The results of this study indicate that air pollutants SO_2, CO and PM_{10} are significantly associated with non-specific conjunctivitis. Air pollutants, SO_2, CO and PM_{10} can increase the chances of outpatient visits, suggesting that this may be the cause of non-specific conjunctivitis.

1. Introduction

Air pollution is associated with many adverse health effects and diseases, and many studies are well aware of this. [1, 2, 3, 4, 5] WHO defines air pollution as "the contamination of the indoor or outdoor environment by any chemical, physical or biological agent that modifies the natural characteristics of the atmosphere. [6]" Although recent legislation on air pollutant standards is more stringent than before, air pollution remains a major problem and its health effects remain. The more developed countries attach great importance to the regulation of air pollution, and the organization and government departments have conducted comprehensive review and guidance.

Previous studies were carried out on the correlation between air pollutants and health using a wide spectrum of mathematical and statistical methods. Findings from most of the previous studies showed that air pollutants have negative influence on health. Studies on lung cancer [1, 2], acute respiratory infection [3], liver cancer [4], and menstrual cycle [5] showed a significant correlation between health and air pollutants. Dense innervations in the ocular surface are extremely sensitive to environmental agents. [7] Furthermore, the eyes are protected only by a thin layer of tear film from potentially damaging exterior influences. [8] Thus, human eyes are very susceptible to ill effects of air pollution.

The ill effects of air pollution on human eyes are mostly irritation and inflammation, with conjunctivitis being a significant problem. [9, 12-15] Many research investigations have attempted to discover the effects of environmental toxins on the ocular surface. Among persons living in high-polluted areas, a considerable portion suffers from subclinical ocular surface changes. [12]

Conjunctivitis is one of the most commonly diagnosed conditions in ophthalmologic outpatient and emergency room visits. [14] More than 40% of ophthalmologic outpatient visits are diagnosed with conjunctivitis annually, according to data from the Bureau of Taiwan National Health Insurance. Conjunctival disorders induced by air pollution can be subclinical ocular surface changes [12], but on many occasions with serious discomforts such as burning and irritation require clinical visits. Furthermore, chronic exposure to air pollution can induce cellular change, such as goblet-cell hyperplasia in the ocular surface. [10] The discomforts of ocular diseases manifested in burning, irritation, itching, and tearing interfere with people's daily activities, including work efficiency and road safety. Treating ocular diseases by steroid eye drops occasionally results in cataracts, glaucoma, and other severe side effects, which can lead to permanent vision loss. [11]

Bourcier, Viboud, Cohen, Thomas, Bury, Cadiot, Mestre, Flahault, Borderie, and Laroche investigated short-term associations between ophthalmologic emergency room visits and air pollution in Paris, and they concluded a relation between NO_2, O_3, SO_2, and PM_{10} and conjunctivitis. [9] Chang, Yang, Chang, and Tsai investigates the relationship between air pollution and outpatient visits for nonspecific conjunctivitis in Taiwan, and they found that the air pollutants, NO_2, SO_2, O_3, and PM_{10} increase the chances of outpatient visits for nonspecific conjunctivitis and have no evident lag effects. [12] Mimura, Ichinose, Yamagami, Fujishima, Kamei, Goto, Takada, and Matsubara examined the association of airborne pollutants with allergic conjunctivitis in Tokyo. Their findings suggest a possible role of $PM_{2.5}$ in the development of allergic conjunctivitis during the non-pollen season. This association between $PM_{2.5}$ and allergic conjunctivitis may have broad public health implications in relation to allergic diseases. [13] Szyszkowicz, Kousha, and Castner examine the associations between emergency department visits for conjunctivitis and ambient air pollution levels in urban regions across the province of Ontario, Canada and they suggest that there are associations between these two, with different temporal trends. [14] Based on the database of the Shanghai Health Insurance System (SHIS), Hong, Zhong, Li, Xu, Ye, Mu, Lu, Mashaghi, Zhou, Tan, Li, Sun, Liu, and Xu assess the relationship between air pollutants with outpatient visits for allergic conjunctivitis. They found that higher levels of ambient NO_2, O_3, and temperature increase the chances of outpatient visits for allergic conjunctivitis. Ambient air pollution may contribute to the worsening of allergic conjunctivitis. [15]

2. Materials and research method

This study used Bayesian linear regression to investigate the relationship between air pollution concentrations and outpatient visits for nonspecific conjunctivitis (Figure 1). In the Bayesian linear regression of the cases of nonspecific conjunctivitis, the SO_2, NO_2, O_2, CO, and PM_{10} are regarded as air pollutants.

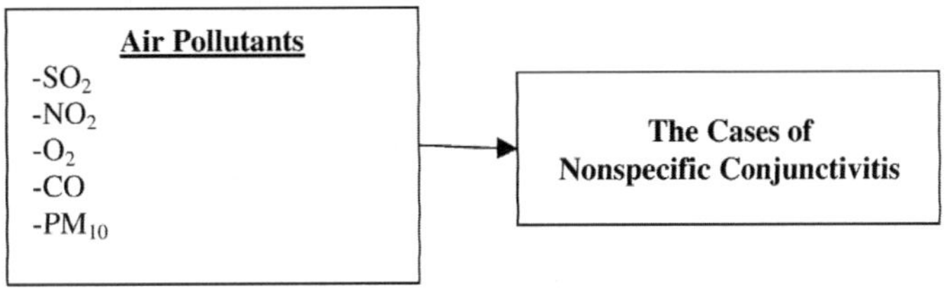

Figure 1. The research framework

2.1. Study area

Taiwan has made efforts on air pollution monitoring and control. The Environmental Protection Administration (EPA) of Taiwan started to set up the Taiwan Air Quality Monitoring Network (TAQMN) in 1990. In September 1993, there were 66 air quality monitoring stations in TAQMN, replacing the original 19. Up until 1998, there were 72 air quality monitoring stations, including four traffic pattern stations, four background pattern stations, three industrial pattern stations and two national park pattern stations. These five types of air quality monitoring stations serve different purposes.

In Taiwan, the priority pollutants that can be monitored through continuous emission monitoring systems (CEMS) of the monitoring network include sulfur dioxides (SO_2), nitrogen dioxide (NO_2), oxygen (O_2), carbon monoxide (CO), and particulate matter (PM_{10}).

2.2. Data collection

The research focuses on the significance of air pollution's impact on ocular health. Data for ophthalmology outpatient visits were obtained from the National Health Insurance Research Database of Taiwan. Nonspecific conjunctivitis was defined and filtered according to the diagnostic codes in the International Classification of Diseases, 9th revision (ICD-9). Ambient air monitoring data were obtained from the Taiwan Environmental Protection Administration air pollution monitoring stations. Six regulated air pollutants were included in this study.

Though many air pollutants are monitored in Taiwan, not every air monitoring station monitors all kinds of the air pollutants because different kind of monitoring stations has its own purpose. This study only concerns about SO_2, NO_2, O_2, CO, and PM_{10}. These air pollutants are regarded as criteria pollutants because the pollutant standard index (PSI) is based on them. Only the monitoring stations that contained data with all these six pollutants were considered for this study.

The clinic visit counts of non-specific conjunctivitis were obtained from the data of National Health Insurance database. This database is originated from the Bureau of National Health Insurance (BNHI) and cleaned by National Health Research Institute (NHRI) of Taiwan. Data from ophthalmologic outpatient visits between 1997 and 2010 were obtained from the National Health Insurance Database. In ICD-9-CM, acute conjunctivitis is coded 372.00; chronic conjunctivitis is coded 372.10; simple chronic conjunctivitis is coded 372.11; other chronic allergic conjunctivitis is coded 372.14; blepharoconjunctivitis is coded 372.20, other and unspecified conjunctivitis is coded 372.30; and conjunctival hemorrhage is coded 372.72. To protect privacy, the data on patient identities and institutions had been scrambled cryptographically.

2.3. Data analysis: Bayesian regression model

The cases of nonspecific conjunctivitis of Taiwan from January 1997 to December 2010 were analyzed with Bayesian linear regression with Julia. [16] The Bayesian linear regression model plots the nonspecific conjunctivitis incidence (cases) versus SO_2, NO_2, O_2, CO, and PM_{10} in the following equation:

$$\text{nonspecific conjunctivitis incidence}=\beta_0+\beta_1*SO_2+\beta_2*NO_2+\beta_3*O_2+\beta_4*CO_2+\beta_5*PM_{10} \quad (1)$$

This study introduced the Bayesian regression model by Martin, Quinn, and Park. The main advantages of the Bayesian approach is that using Markov chain Monte Carlo (MCMC) to output backward reasoning is very simple, as these methods provide a direct sample of the parameters of interest. [17-19] In purpose of to detect the significance of air pollution's impact on ocular health, Bayesian regression model be applied to insurance data. This model [17-19] simulates from the posterior distribution using standard Gibbs sampling which is a multivariate Normal draw for the betas, and an inverse Gamma draw for the conditional error variance. The model to be estimated is as:

$$Yi=x'_i\beta+\varepsilon_i \quad (2)$$

Where the errors are assumed to be Gaussian as $\varepsilon_i \sim N(0,\sigma^2)$. It can be assumed standard, semi-conjugate priors as $\beta \sim N(b_0, B_0^{-1})$, and $\sigma\text{-}2 \sim Gamma(c_0/2,d_0/2)$.

Where β and σ^{-2} are assumed a priori independent. Note that only the starting value of β is allowed because the Gibbs sample with the conditional error variance is used for the simulation as the first block in the sampler.

3. Results

Using monthly aggregated data, outpatient visit data of nonspecific conjunctivitis of Taiwan from 1997 to 2010 were analyzed with Bayesian linear regression analysis for the interval-valued data in this study. The air pollutants data in this study includes SO_2, NO_2, O_2, CO, and PM_{10}. Details of the non-specific conjunctivitis incidence in Taiwan, SO_2, NO_2, O_2, CO, and PM_{10} measurements are presented in Table 1.

Table 1. The Summary of the research data.

	N	Min	Max	Average	S.E.
SO_2	168	2.41	7.14	4.563	0.895
NO_2	168	11.51	27.81	19.295	4.277
O_2	168	16.20	43.80	27.568	5.478
CO	168	0.32	0.92	0.614	0.142
PM_{10}	168	32.00	96.00	57.790	15.130
The Cases of Nonspecific Conjunctivitis	168	888.00	59725.00	29778.200	14016.945

Table 2. The result of Bayesian linear regression analysis of air pollutants affecting nonspecific conjunctivitis cases in Taiwan.

	Mean	St. Dev	Naive SE	Lower	Upper
(Intercept)	73461.7	6187.4	61.874	61451.7	85566.5
SO_2	6716.7	1064.5	10.645	4631.5	8798.5
NO_2	257.7	507.6	5.076	-730.8	1243.5
O_2	125.9	171.6	1.716	-212.0	463.6
CO	67757.4	11062.1	110.621	46386.4	89271.9
PM_{10}	347.6	102.7	1.027	148.4	551.4
σ^2	71913692.0	8106157.7	81061.577	57613085.1	89527014.2

The result of Bayesian linear regression analysis is as Table 2. Lower and Upper indicate central (IC) 95 percent Bayesian credible intervals in Table2. SO2 (95% IC: 4631.5~8798.5), CO (95% IC: 46386.4~89271.9), and PM10 (95% IC: 148.4~551.4) are positively correlated with nonspecific conjunctivitis cases. The relationship between nonspecific conjunctivitis cases and NO2 (95% IC: -730.8~1243.5) and O2 (95% IC: -212.0~463.6) are not statistically significant.

4. Discussion and conclusions

Previous studies examining the relationship between air pollution and respiratory diseases have shown that air pollution can induce and aggravate respiratory diseases. [20] Some studies suggest that the response of the conjunctival and respiratory mucosa is similar to exogenous stimuli. [21] The results of this study show some similar patterns consistent with respiratory disease investigations.

In this study, the results showed that air pollutants, SO_2, CO and PM_{10} were significantly associated with non-specific conjunctivitis. The more the increase of the level of SO_2, CO, and PM_{10}, the more nonspecific conjunctivitis cases occur. This result is similar with the studies of Bourcier, Viboud, Cohen, Thomas, Bury, Cadiot, Mestre, Flahault, Borderie, and Laroche [9], Chang, Yang, Chang, and Tsai [12], Szyszkowicz, Kousha, and Castner [14], and Hong, Zhong, Li, Xu, Ye, Mu, Lu, Mashaghi, Zhou, Tan, Li, Sun, Liu, and Xu. [15]

The study combines and integrates air pollution and ophthalmology data to investigate the association between non-specific conjunctivitis clinic visits and air pollution levels. It has found that the air pollutant, SO_2, CO, and PM_{10} can increase chances of outpatient visits, suggesting that it is a possible cause for nonspecific conjunctivitis. Furthermore, this paper has also shown that a favourable air-pollution-monitoring infrastructure combined with established public health data can provide valuable information for an examination of adverse health effects related to air pollution.

References

IOP Conf. Series: Materials Science and Engineering **658** (2019) 012017 doi:10.1088/1757-899X/658/1/012017

[1] S Yue, Y Wang, J Wang and J Chen 2017 *Technology and Health Care* **25**(S1) pp 411 – 422
[2] M Dehghani, L Keshtgar, M R Javaheri, Z Derakhshan, G O Conti, P Zuccarello and M Ferrante 2017 *Molecular Medicine Reports* **15**(5) pp 3390 – 3397
[3] R Li, N Jiang, Q Liu, J Huang, X Guo, F Liu and Z Gao 2017 *International Journal of Environmental Research and Public Health* **14**(1) http://www.mdpi.com/1660-4601/14/1/47
[4] H Deng, S P Eckel, L Liu, F W Lurmann, M G Cockburn and F D Gilliland 2017 *International Journal of Cancer* **141**(4) pp 744 – 749
[5] A Merklinger-Gruchala, G Jasienska and M Kapiszewska 2017 *International Journal of Environmental Research and Public Health* **14**(7) http://www.mdpi.com/1660-4601/14/7/816
[6] Word Health Organization: Air pollution https://www.who.int/airpollution/en/
[7] I S J Tuominen, Y T Konttinen, M H Vesaluoma, J A O Moilanen, M Helinto and T M T Tervo 2003 *Investigative Ophthalmology & Visual Science* **44**(6) pp 2545 – 2549
[8] J Wang, D Fonn, T L Simpson and L Jones 2003 *Investigative Ophthalmology & Visual Science* **44**(6) pp 2524 – 2528
[9] T Bourcier, C Viboud, J C Cohen, F Thomas, T Bury, L Cadiot, O Mestre, A Flahault, V Borderie and L Laroche 2003 *The British Journal of Ophthalmology* **87**(7) pp 809 – 811
[10] P Novaes, P H do Nascimento Saldiva, N Kara-José, M Macchione, M Matsuda, L Racca and A Berra 2007 *Environmental Health Perspectives* **115**(12) pp 1753 – 1756
[11] R Gerometta, S M Podos, J Danias and O A Candia 2009 *Investigative Ophthalmology & Visual Science* **50**(2) pp 669 – 673
[12] C J Chang, H H Yang, C A Chang and H Y Tsai 2012 *Clinical and Epidemiologic Research* **53**(1) pp 429 – 433
[13] T Mimura, T Ichinose, S Yamagami, H Fujishima, Y Kamei, M Goto, S Takada and M Matsubara 2014 *Science of the Total Environment* **487** pp 493 – 499
[14] M Szyszkowicz, T Kousha and J Castner 2016 *International Journal of Occupational Medicine and Environmental Health* **29**(3) pp 381 – 393
[15] J Hong, T Zhong, H Li, J Xu, X Ye, Z Mu, Y Lu, A Mashaghi, Y Zhou, M Tan, Q. Li, X Sun, Z Liu and J Xu 2016 *Scientific Reports* **6** https://www.nature.com/articles/srep23858.pdf
[16] The Julia Programming Language https://julialang.org/
[17] A D Martin, K M Quinn and J H Park 2011 *Journal of Statistical Software* **42**(9) pp 1 – 21 http://www.jstatsoft.org/v42/i09/
[18] S Chib 1995 *Journal of the American Statistical Association* **90** pp 1313 – 1321
[19] R E Kass, A E Raftery 1995 *Journal of the American Statistical Association* **90** pp 773 – 795
[20] M Brauer, G Hoek, P Van Vliet, K Meliefste, P H Fischer, A Wijga, L P Koopman, H J Neijens, J Gerritsen, M Kerkhof, J Heinrich, T Bellander and B Brunekreef 2002 *American Journal of Respiratory and Critical Care Medicine* **166**(8) pp 1092 – 98
[21] L Bielory and M H Friedlaender 2008 *Immunology and Allergy Clinics of North America* **28**(1) pp 43 – 58

IOP Conf. Series: Materials Science and Engineering **658** (2019) 012018 doi:10.1088/1757-899X/658/1/012018

The dynamic characteristic of aquaculture net cage considering flexible twine damaged

Zhongqiang Zheng[1], Rong Wan[2,3], Zongyu Chang[1,4] and Mingtao Chi[1]

[1]College of Engineering, Ocean University of China, Qingdao, China
[2]College of Marine Science, Shanghai Ocean University, Shanghai, China
[3]National Engineering Research Center for Oceanic Fisheries, Shanghai, China
[4]Key Laboratory Ocean Engineering Lab of Shandong Province, Qingdao, China

E-mail: zqzheng@ouc.edu.cn

Abstract. The aquaculture net cage can provide the living space for fish in larger waves and stronger currents. However, if the fish escapes from the damaged of aquaculture net cage, the wild fish populations will be threated. In this paper, the dynamic model of the circular net was developed in steady current using lumped mass method. In order to validate the numerical model, the deformations of netting were compared with the physical model test and the numerical results by other authors. Then the dynamic result of flexible twine damaged could be obtained by numerical simulation analysis in current. The variety of netting deformation, the tension distribution of the circular net and the force of netting twine connected the floating collar were analyzed. The result show that the flexible twine damaged will lead to the local variable. This study is helpful for the predication of flexible twine damaged through measure the force distribution of floating collar.

1. Introduction

Aquaculture farming is playing an important role in the future, and can provide more and more the marine food for human consumption. There is a trend of moving marine fish farms to more exposed areas. The aquaculture net cage must face the larger waves and stronger currents. So the aquaculture net cage will faced with huge threat. Once aquaculture net cage is damaged or collapsed leading to fish escaping. The wild fish populations will be threated and the fish farmers have economic losses. Thence the predication of flexible twine damaged can through measure the force distribution of floating collar.

Many scholars have studied the hydrodynamic characteristics of net [1-4]. In order to making the established models more realistic, the numerical models of the net were optimized through compared with the test data. Lee et al. [5] calculated the dynamic behavior of net using implicit integration method. Huang et al. [6, 7]studied the dynamic response of the gravity cages by the lumped-mass method in currents. Cifuentes et al. [8] designed the complex cage system using Morison force model with variations Cd according to high solidity ratio and current speed. Moe-Føre et al. [9] established the numerical model for high solidity net cage at high flow velocities considering the changes in water flow and the global drag coefficient of the net. Ma et al. [10] proposed an innovative hybrid method to simulate the interaction between net and water flow. Yao et al. [11] used novel hybrid volume approach to establish the calculation model of net under the current load.

The research subjects are focus on the hydrodynamic characteristics of net cage. According to the statistical analysis, which two-thirds of the escape incidents and number of escaping fish are due to hole in the net, damage of net is the main reason for the escape of fish [12]. So this paper will developed the dynamic model of aquaculture net cage considering flexible twine damaged. Through compared with the physical model test and the numerical results by other authors, the results of numerical model are compared in this paper. The netting deformation, the tension distribution of the circular net and the force of netting twine connected the floating collar are obtained using numerical simulation analysis.

2. Numerical model

In this paper, the flexible net was simplified mass-spring model using lumped mass method as shown in Figure 1. Only considering the steady current, the forces of each knot can be expressed in the Figure 1 including the gravitational force $\boldsymbol{W}$, buoyant force $\boldsymbol{B}$, drag force $\boldsymbol{F}$ and tension force $\boldsymbol{T}$. Therefore, the motion equation of circular net can be described as:

$$M\frac{\partial\overline{q(t)}}{\partial t}=\overline{W(t)}+\overline{B(t)}+\overline{F(t)}+\overline{T(t)} \tag{1}$$

Where, M is the mass of the mesh knot, $\overrightarrow{q(t)}$ is the velocity of mesh knot.

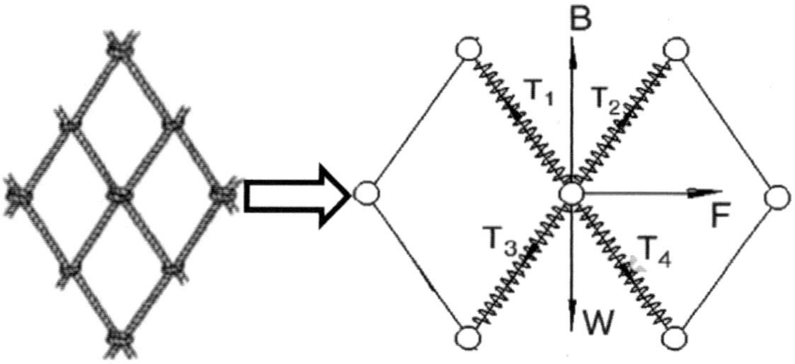

Figure 1. The force of mesh knot in mass-spring model of flexible net

The gravitational force of mesh knots:

$$W=m_i g \tag{2}$$

where, g is the acceleration of gravity. The value in this paper is $9.81 m/s^2$.

The buoyant force of mesh knots:

$$B_i=\rho g \forall_i \tag{3}$$

where, ρ is the density of the water; $\forall_i$ is the volume of water that the node discharges.

The tension between node i and adjacent node j (j=1, 2, 3, 4) can be calculated as following

$$T_{ij}=\begin{cases} AC_1\left(\dfrac{l_{ij}-l_0}{l_0}\right)^{C_2} & l_{ij}>0 \\ 0 & l_{ij}\le 0 \end{cases} \tag{4}$$

where, l_0 is the original length of among knots, l_{ij} is the length after deformation,i.e. $l_{ij}=\sqrt{(x_j-x_i)^2+(y_j-y_i)^2+(z_j-z_i)^2}\quad (i,j=1,2,3,4)$, A is the sectional area of the mesh knots, and C_1 and C_2 are the elastic coefficients of the flexible net material. According to Fibre Ropes for Fishing

IOP Conf. Series: Materials Science and Engineering **658** (2019) 012018 doi:10.1088/1757-899X/658/1/012018

Gear by Klust [13], the elastic coefficients of polyethylene (PE) material are given as C_1=3.454×10^8 and C_2=1.0121.

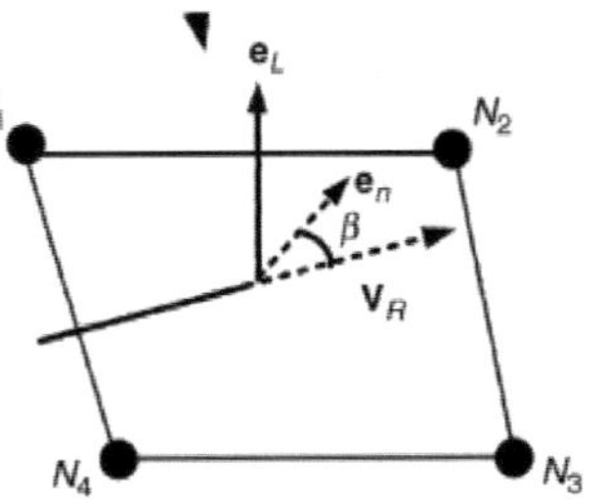

Figure 2. Relation between a flexible net cage and fluid flow

The current is assumed to be steady and uniform, so the inertial force is neglected. Figure 2 shows the relationship between a flexible net cage and steady current. Assuming a water particle is moving into the net cage with a relative velocity $\boldsymbol{U_R}$ to the net panel. The drag force and lift force can be expressed as:

$$\begin{cases} F_{Di} = -\frac{1}{2}\rho C_D A_p U_R |U_R| \overline{n_D} \\ \overline{F_{Li}} = \frac{1}{2}\rho C_L A_p U_R |U_R| \overline{n_L} \end{cases} \quad (5)$$

where, C_D is the drag force coefficient, A_P is the projected area of the mesh knots, U_R is the relative velocity of the fluid flow and the mesh knots, C_L is the lift force coefficient.

3. Numerical simulations and results

In order to verify the accuracy of the numerical calculation model, the numerical results were compared with the data of physicals experimental model by Lader and Enerhaug [14]. The physicals experiment net was made up of two rectangular net. Each rectangular net was 81 meshes high and 125 meshes wide. The material of the flexible net was nylon (PA). The grid had a side length of 16*mm* and the twine thickness was 1.8*mm*. The diameter of the circular net was 1.435*m* and its height was 1.44*m*. The weight was composed of 16 weights evenly distributed at the bottom of the circular net and the mass of each weight has 400*g*. The current velocities of test were 0.21*m/s* and 0.33*m/s*, respectively. Figure 3(a) shows the experimental results by Lader and Enerhaug [14], Figure 3(b) shows the numerical results by Zhao et al. [15], Figure 3(c) show the numerical results of this paper. From Figure 3 and Table 1, it can be seen that the numerical results agreed well with the experimental datas, indicating that the numerical calculation method used in this paper is reasonable.

In this paper, the shape of element in circular net is diamond. The diameter of floating collar at the top of the net is 1.435*m* and 32 weights are evenly distributed at the bottom of the net with fixed [16]. The mass of each weight is 200*g*. The different current velocity (0.22*m/s*, 0.33 *m/s*, and 0.44*m/s*) were used to impact the flexible net, respectively. The other parameters using in simulation are listed in Table 2.

Table 1. Comparison of volume reduction rate and drag force on the net between calculated and experimental results

Velocity/ (*m/s*)	**volume reduction rate**			**drag force/N**		
	Experiment/ %	**Calculation/ %**	**Absolute error/%**	**Experiment**	**Calculation**	**Absolute error/%**
0.13	3	1	2	12.02	11.27	6%
0.21	9	5	4	21.97	23.4	7%
0.26	14	11	3	34.96	35.88	3%
0.33	23	24	1	49.79	57.82	16%

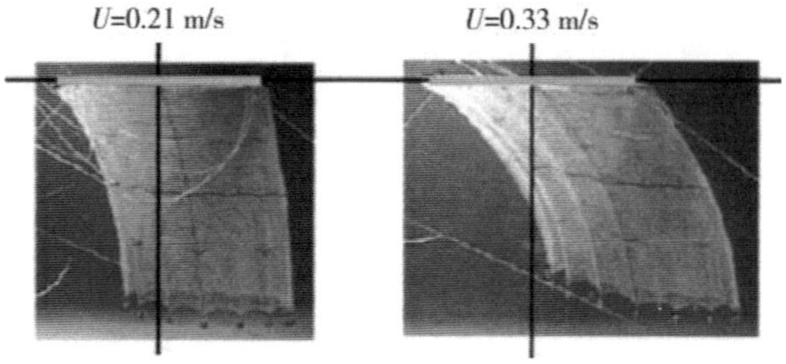

(a) Experimental results (Lader and Enerhaug, 2005)

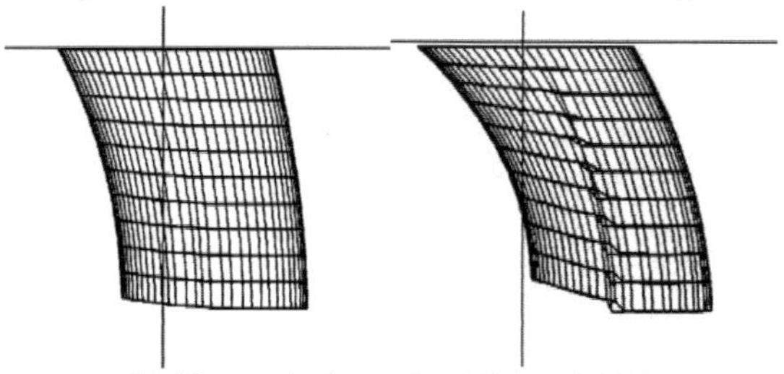

(b) Numerical results (Zhao, 2007)

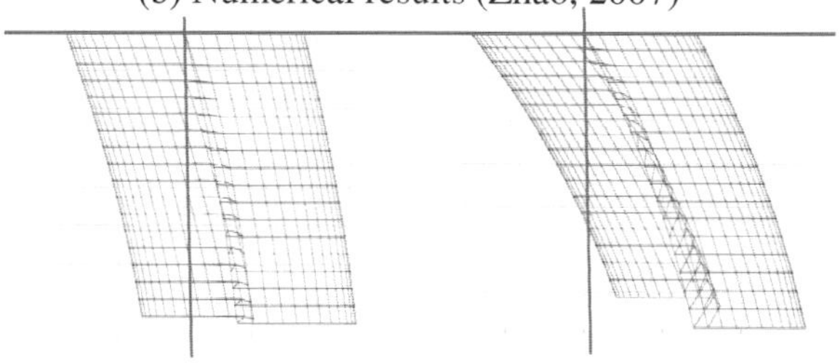

(c) Numerical results of this paper

Figure 3. The deformation of circular net in different steady current

Table 2. The parameters using in simulation

Item	Material	Height of net(*m*)	The density of PE(*kg/m*^{3})	Mesh size(*mm*)	Diameter of twine(*mm*)	density of water(*kg/m*^{3})
Condition	PE	1.44	953	45	1.8	1025

The circular net has so many nodes, which leads to inefficient computing. So the neighboring 5x5 mesh is equivalent to a large mesh using the mesh grouping method [17]. The structural model is shown in Figure 4(a) and the flexible twine damaged model is shown in Figure 4(b).

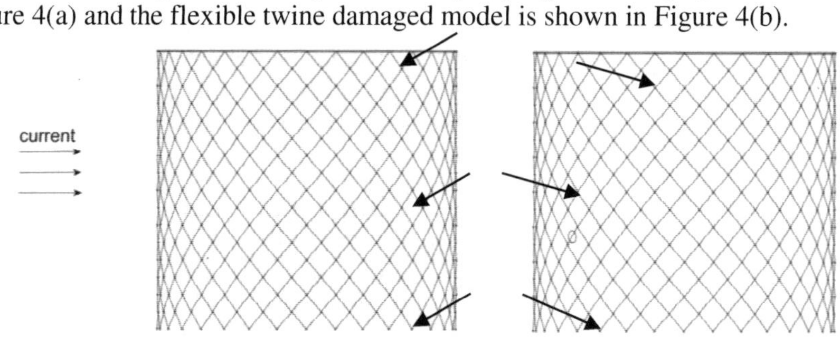

(a) (b)

Figure 4. Structure of circular net

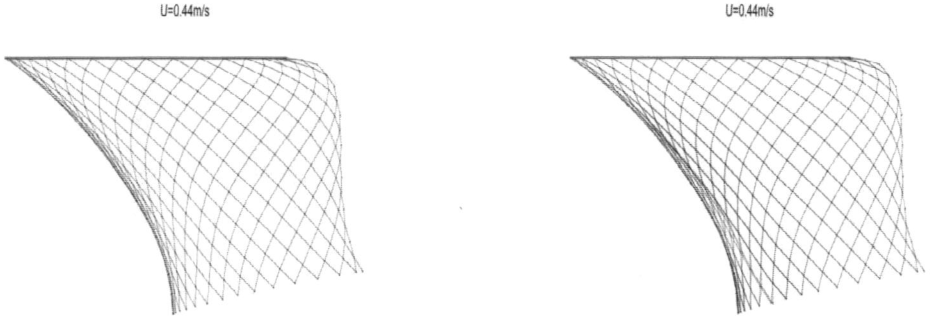

(a) The deformation of undamaged circular net (b) The deformation of damaged circular net

Figure 5. The steady shape of circular net in current

From Figure 5, there is not much difference in the deformation of undamaged and damaged flexible net. Because the flexible net is round, it is inconvenient to observe tension distribution on netting. Therefore, from the demarcation 1 between the upstream net and the downstream net, it is unfolded in a counterclockwise direction to obtain a plane expanding view of the netting forces, which were shown in Figure 6.

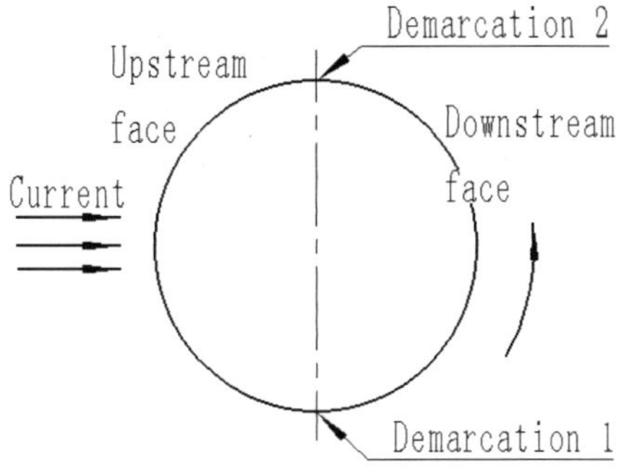

Figure 6. The unfolded method of circular net (top view)

The right part of Figure 7 shows tension distribution at the upstream net and the left part shows tension distribution at the downstream net. The forces on the upstream net and the downstream net of undamaged circular net are symmetrically distributed. If the twine is damaged, the force in the damaged area will change. The tension force of the damaged diagonal twine is reduced, but the tension force of the adjacent diagonal line parallel is increased. The other forces are almost unchanged. The result is shown in Figure 8.

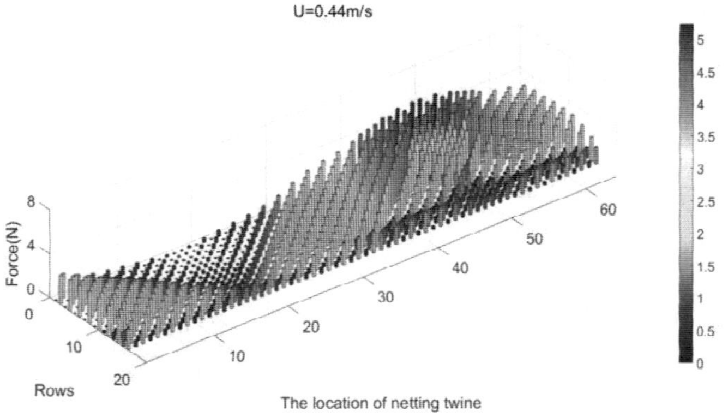

(a) The tension distribution of undamaged circular net

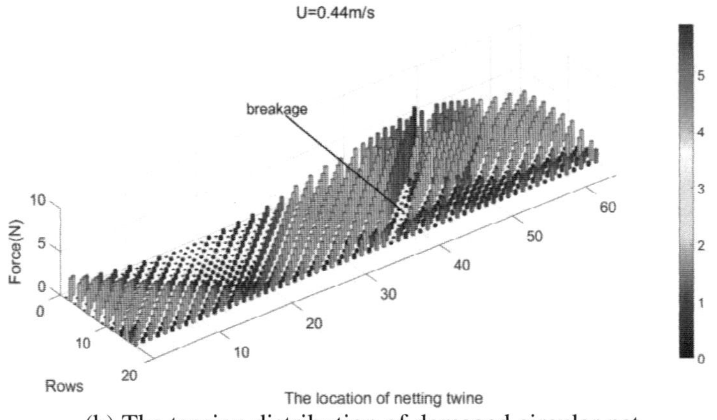

(b) The tension distribution of damaged circular net

Figure 7. Tension distribution of circular net

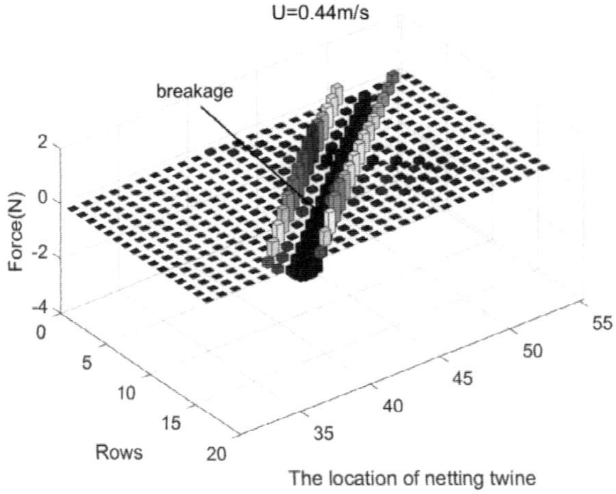

Figure 8. Difference between condition of netting with undamaged and damaged (Partial enlargement view)

Tension forces distribution of twine on floating collar are shown in Figure 9. Under normal circumstances, the force on the net is not easy to obtain. Therefore, we can use the tension force on the floating collar to predict the change of force on the net. From Figure 9, the oblique direction in which the tension is reduced is the direction of the damage. According to the direction, the damaged circular net can be found for reduce losses.

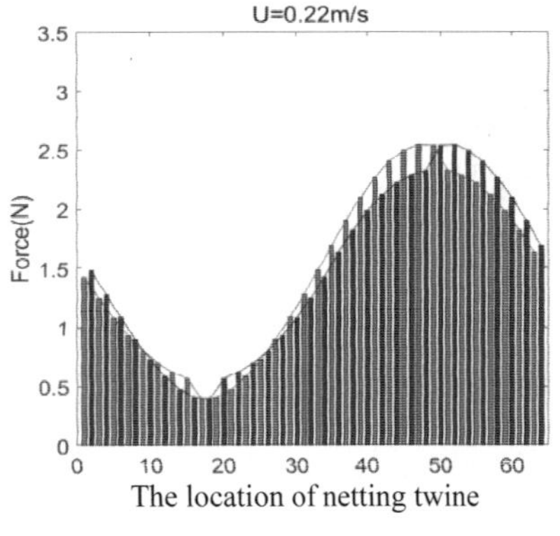

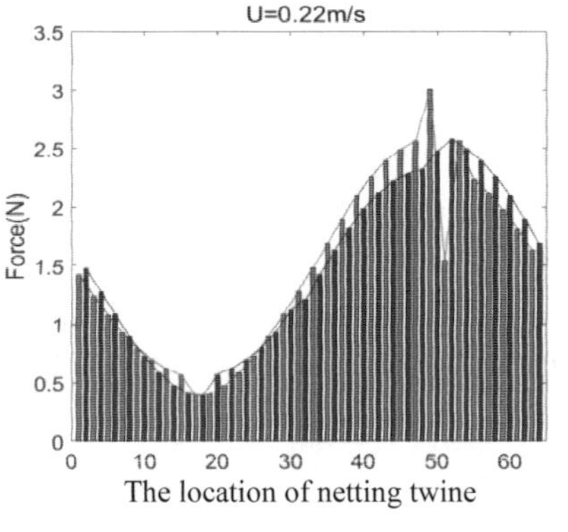

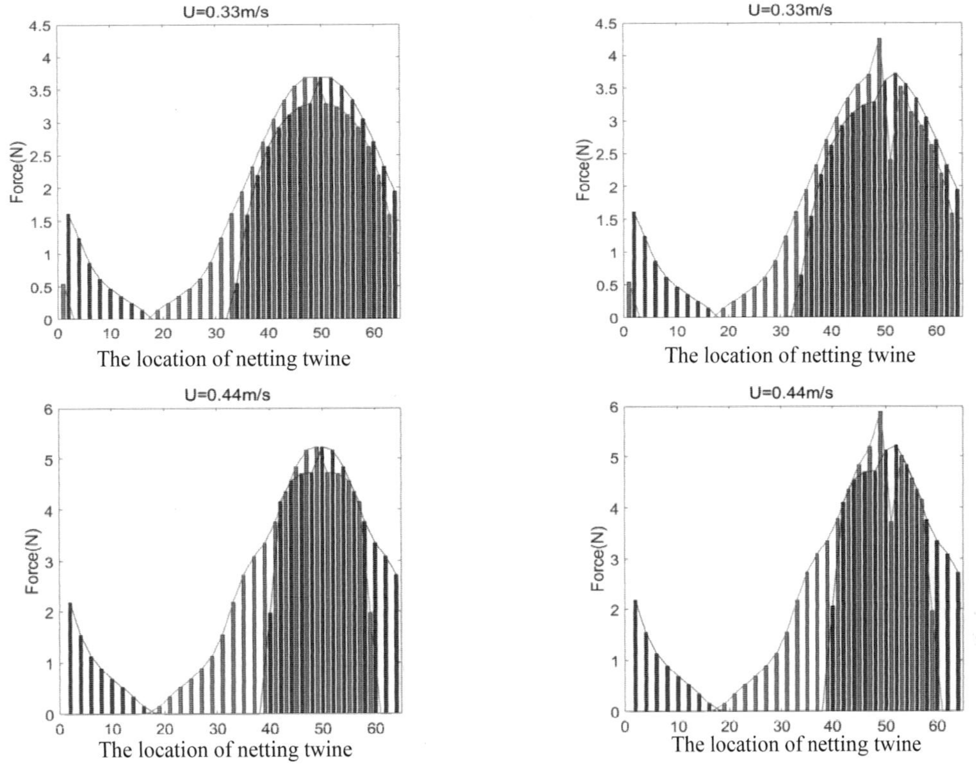

(a) undamaged circular net (b) damaged circular net

Figure 9. The force characteristic of floating collar

4. Conclusion

The dynamic characteristic of circular net with twine damaged in net cage under steady current is studied. The mathematical model of the system is established using lumped mass method. The shape of circular net, tension force distribution on the twine and floating collar are obtained by numerical analysis. Single twine damaged has little effect on deformation of circular net. The tension forces of twine are changed on the local area of twine damaged especially adjacent diagonals. The tension forces of floating collar are symmetrical without damaged, but the tension forces would be changed with damaged. The oblique direction in which the tension is reduced is the direction of the damage, and adjacent diagonals direction the forces is increased. So the forces can be measured by sensor in order to judge if the twine is damaged. The work is very useful for predict the damage direction of circular net cage.

Acknowledgements

The research was supported by the National Natural Science Foundation of China (Grant No.51175484), the Fundamental Research Funds for the Central Universities (Grant No.201513053) and the Key Laboratory for Mechanism Theory and Equipment Design of Ministry of Education (Tianjin University).

References

[1] R Wan, F Hu and T Tokai 2010 *Fisheries Sci.* **68**(4) pp 815 – 823

[2] Tsukrov I, Eroshkin O, Fredriksson D, Swift M R and Celikkol B 2003 *Ocean Eng.* **30** pp 251 – 270

[3] Zhan JM, Jia XP, Li YS, Sun MG, Guo GX and Hu YZ 2006 *Aquacult Eng.* **35** pp 91 – 101

IOP Conf. Series: Materials Science and Engineering **658** (2019) 012018 doi:10.1088/1757-899X/658/1/012018

[4] Zhao YP, Gui FK, Xu TJ, Chen XF and Cui Y 2013 *Appl. Ocean Res.* **39** pp 158 – 167
[5] Lee CW, Lee JH, Cha BJ, Kim HY and Lee JH 2005 *Ocean Eng.* **32** pp 331 – 347
[6] Huang CC, Tang HJ and Liu JY 2006 *Aquacult. Eng.* **35** pp 258 – 270
[7] Huang CC, Tang HJ and Liu JY 2007 *Aquacult. Eng.* **37** pp 144 – 157
[8] Cifuentes C and Kim MH 2017 *Ocean Systems Eng.* **7**(2) pp 143 – 155
[9] Moe-Føre H, Lader PF, LienO E and Hopperstad S 2016 *J. Fluid. Struct.* **65** pp 180 – 195
[10] Ma L, Hu K, Fu S, Moan T and Li R 2016 *J. Ship Res.* **60** pp 14 – 29
[11] Yao YM, Chen YL, Zhou H and Yang HY 2016 *J. Fluid. Struct.* **62** pp 350 – 366
[12] Jensen O, Dempster T, Thorstad EB, Uglem I and Fredheim A 2010 *Aquacult. Env. Interac.* **1** pp 71 – 83
[13] Klust G 1983 *Fibre ropes for fishing gear* (Madras, India)
[14] Lader PF and Enerhaug B 2005 *IEEE J. Oceanic Eng.* **30**(1) pp 79 – 84
[15] Zhao YP, Li YC, Dong GH, Gui FK and Teng B 2007 *Aquacult. Eng.* **36** pp 285 – 301
[16] Huang XH, Guo GX, Hu Y and Tao QY 2010 *J. Fishery Sci. China* **17** pp 312 – 319 (in Chinese)
[17] Su W and Zhan JM 2007 *J. Hydrodyn.* **22** pp 267 – 272 (in Chinese)

IOP Conf. Series: Materials Science and Engineering **658** (2019) 012019 doi:10.1088/1757-899X/658/1/012019

An invention of multifunction washing ball

Che Wei Liu and Chuan Wei Wu

Department of Civil Engineering National Taipei University of Technology, Taiwan

E-mail: Liou6@seed.net.tw

Abstract. Ultrasonic clean is a very fashionable trend in cleaning appliance and tools in domestic and international markets, from washing machine, electronic toothbrush to household, ultrasonic cleaning machine (for jewelry, watch, glasses, etc.) Every ultrasonic related thing is very popular and the price is raised immediately. For example, the washing machines under the same brand, the one with ultrasonic will be USD 180 more expensive than the other one as the price could up to 8 times of difference for the ultrasonic function of the electronic toothbrushes of the same brand. Multifunction washing ball is the pioneer creation of the world, which is advanced and can be applied for many purposes. It can replace various cleaning household appliances in modern family with the remarkable cleaning effect. Ultrasonic current powerfully get rid of the hidden dirt among the clothes fiber. It will definitely be popular with the collocation of the trend. In the environmental protection, it reduces washing powder usage, minimum power consumption, save the water, cherish the resource on earth.

1. Introduction

Human can hear the sound frequency from 16Hz to 20 KHz, ultrasonic means the frequency that human can not to hear. Ultrasonic cleaning uses cavitation bubbles induced by high frequency pressure waves to agitate a liquid[1]. The agitation produces high forces on contaminants adhering to substrates like dust, dirt, oil, pigments, rust, grease, algae, bacteria, polishing compounds, soot wax, and so on[2].

Ultrasonic cleaning machine is conversed to the mechanical vibration with the same frequency through the piezoelectric inverse-transforming effect of the transducer, and then it will radiate forward intensively in the cleaning solution in the form of ultra-audio-frequency longitudinal wave (The ultra-audio frequency longitudinal wave spreads alternatively between zones of negative pressure and positive pressure), producing ten thousands of small bubbles to explode in turn. These small bubbles which form and grow in the zone of negative pressure, and explode instantaneously in the zone of positive pressure[3], to strike with high pressure the surface, gap, and small local of a cavity to be washed, leading to peeling off of fouling quickly[4].

2. Method

2.1. Past technique

Clothing is generally washed in machines nowadays. However, some articles such as women's undergarments and infants clothes, do not need to go into a washing machine and some may even be damaged if this were done. Hand washing can be inconvenient and some people may not like to send their washing to a hotel laundry when they are traveling. It is also not a good idea to use detergents for

IOP Conf. Series: Materials Science and Engineering **658** (2019) 012019 doi:10.1088/1757-899X/658/1/012019

infants clothing because they can cause asthma, allergy and irritation to the infants skin. That is why pediatricians do not recommend using them for washing baby cloths [5].

Furthermore, the use of copious quantities of water to remove the large amount of detergents needed for very dirty washing can stress the biological habitat and is environmentally unfriendly.

2.2. TRIZ theory

TRIZ is the abbreviation of the Russian "Teoriya Resheniya Izobretatelskikh Zadath", translated into English as the Theory of Inventive Problem Solving (TIPS), [6] which means the theory of innovative problem solving. This theory aims to solve the problems during innovative invention.

TRIZ was developed by Soviet patent researcher Genrish Squlovich Altshuller (1926-1998) from the analysis of the basic concept of patent documents. He found that various innovative invention problems in different fields had common basic problems and problem-solving skills; hence, he proposed the same solution to be used in the problems in different fields and different periods. From these patents, various problem models were systematically sorted out, and he proposed the approach of innovative invention and its problems-solving methods. [7] The theory of TRIZ was published in 1959.

The hypothesis assumed that "there exists some universal invention principles which can be used as the basis for the inventing of new advanced skills. If these invention principles can be confirmed and organized to teach the inventors, then the invention process will be more predictable." The study was divided into several phases and lasted for 50 years. [8] More than one million patent documents were analyzed, and the patents were graded according to the difficulty of the invention. Three findings were obtained:

1. Repeated use of the standard questions and standard solutions in both industry and the scientific community
2. Repeatedly occurring types of technological evolution in both industry and the scientific community
3. Ability to facilitate technological innovation by using the scientific effects outside the field of interest

As a result, he has proposed that the inventive method theory must satisfy the following conditions:

1. It should be a systematic, repetitive, and reliable problem-solving procedure
2. It can guide the inventor to find the ideal solution from various answers
3. It is a method that can increase the invention knowledge
4. It is a structured and convenient system that enables the inventors to overcome the psychological barriers to enhance creativity.

TRIZ is a problem-solving method. It analyzes the problem and finds the contradiction. The contradiction is divided into Physics Contradiction and Technique Contradiction. [9] Different solutions are taken for each contradiction. The theory contains a series of innovative methods, but the most commonly used tool is the Contradictions Matrix and 40 Inventive Principles derived by Altshuller. The application of TRIZ has gradually been valued by Western countries in recent years and caused extensive discussion worldwide[10].

Mechanical engineers usually encounter conflicting issues when they solve engineering problems. Solving one of the problems will result in another problem, which implies that improving one engineering characteristic of the system will result in the deterioration of another engineering characteristic. The TRIZ method can effectively help engineers to solve engineering system interactive problems. To solve the problem, first, the Contradiction Characteristics of the system must be matched to Altshuller's engineering parameters. Then, one finds the relevant inventive principles to solve the problem using the Contradictions matrix flowchart as shown in Figure 1.

IOP Conf. Series: Materials Science and Engineering **658** (2019) 012019 doi:10.1088/1757-899X/658/1/012019

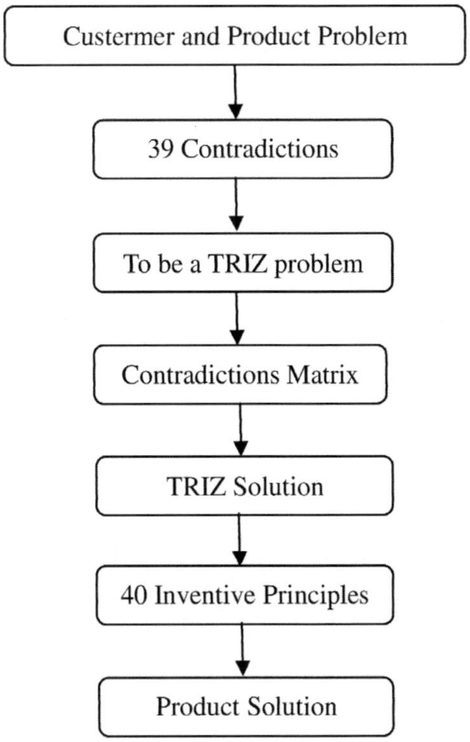

Figure 1. Flowchart

2.3. Improvement

A washing ball with ultrasonic function is disclosed here. The washing ball may be vibrated by using a motor. The washing ball includes a shell body with at least one wing part formed thereon. The motor is configured into the shell body and has an off-centered component. A battery chamber is used for accommodating a dry battery to provide power for the motor.

This type of new multifunction washing ball utilizes ultrasonic technology for cleaning and is completely waterproof. This ball can be placed inside a washing machine to effectively turn it into an ultrasonic washer[11]. The ultrasonic washing ball can reduce the amount of detergent needed by half and still achieve complete cleaning of the cloth. No detergent is needed for washing infants clothes, just clean water. The strong ultrasonic agitation removes hidden dirt from the fabric fibers. This washing ball can upgrade and turn a traditional washing machine into an ultrasonic washer. The washing ball can also be used without a machine. When traveling, just place the washing ball in a wash basin with some water and your intimate clothing. There will be no need for any hand scrubbing. The multifunction washing ball will do the job and is clean and environmentally friendly.

3. Description

In this study, we developed a washing ball with a main shell. The outer part of the shell is equipped with wings along different directions. The shell is housed in left and right shells, which can be screwed together. A circular rubber gasket is arranged between the shells to prevent water from entering the shell. A battery-powered ultrasound device is arranged inside the shell. When the two shells are screwed together, the device generates ultrasonic waves in the water to improve the cleaning effect through the conductive sheet of the battery. The shell g is made of ethylene-vinyl acetate (EVA) and plastic. The wings and shells are formed integrally. The shell can be spherical or elliptical. A semi-circular zigzag structure is designed at the edges of the wings to mimic rubbing during cleaning.

IOP Conf. Series: Materials Science and Engineering **658** (2019) 012019 doi:10.1088/1757-899X/658/1/012019

The wings have a few holes. The inner edges of the holes have fine fibers to assist with the generation of bubbles.

The washing ball described herein is slightly larger than a conventional laundry ball. It consists of a shell 13 with wings 2 in different directions. The shell 13 and the wings 2 are formed integrally by using EVA and plastic or rubber. The shell 13 can be circular, elliptical, or arbitrarily shaped. The edges of the wings 2 are arc-shaped to increase the effect of rubbing during cleaning. A few holes 3 are distributed on the wings 2. The inner edges of the holes have fine fibers 17 to guide the flow of water and assist with the generation of bubbles to enhance the cleaning effect. Figure 2. shows the appearance of the multifunction washing ball.

Figure 3 shows a cross-sectional view of the developed washing ball, including the left shell 20 and right shell (21), wings in the outer portion 2, holes 3 distributed on the wings with fine fibers 17 to guide the flow of water and assist with the generation of bubbles. The inner part houses the ultrasonic device 7 and the battery chamber 11. The ultrasonic device 7 and the battery chamber 11 are arranged inside the left shell 20 and the right shell 21, respectively. The battery inside battery chamber 11 is used as a power supply. The actual application is shown in Figure 4.

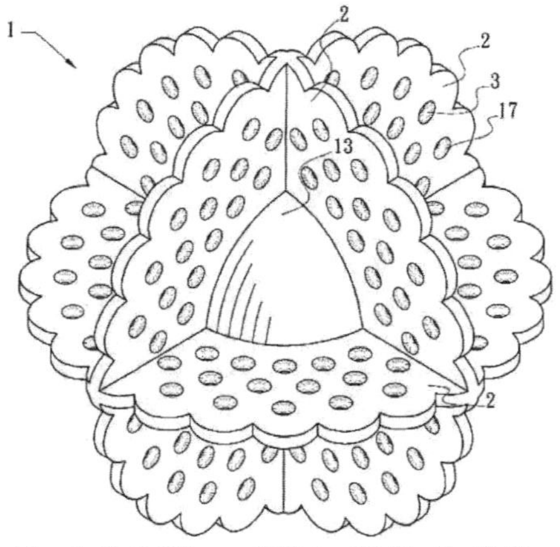

1 Washing ball, 2 Wings, 3 Hole, 13 Shell, 17 Fine fiber

Figure 2. Appearance of Multifunction Washing Ball

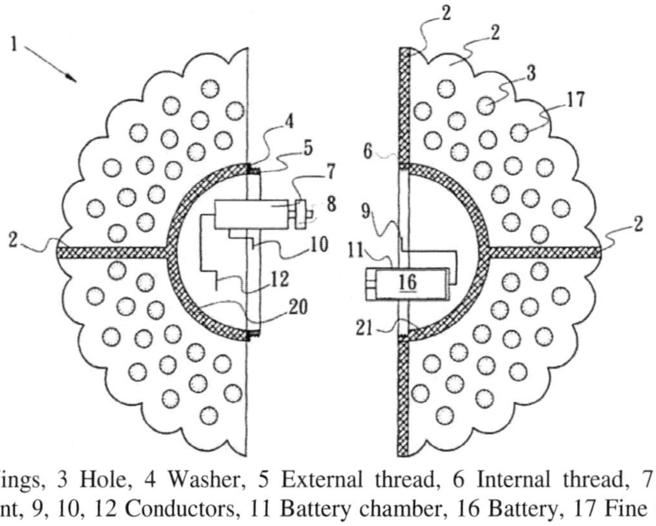

1 Washing ball, 2 Wings, 3 Hole, 4 Washer, 5 External thread, 6 Internal thread, 7 Ultrasound device, 8 Off-centered component, 9, 10, 12 Conductors, 11 Battery chamber, 16 Battery, 17 Fine fiber, 20 Left shell, 21 Right shell.

Figure 3. Component of Multifunction Washing Ball

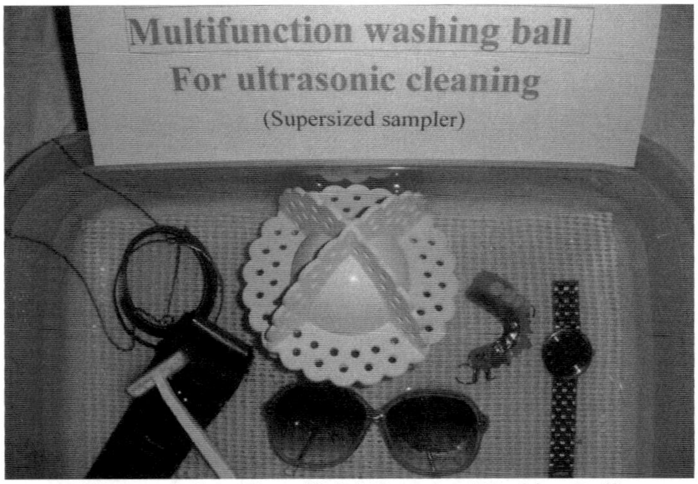

Figure 4. Sampler of Multifunction Washing Ball

4. Conclusions

The multifunction ultrasonic washing ball has already been granted a patent as well as many awards. This is a brand new type of cleaning method. The strong ultrasonic cleaning power reduces the amount of detergent required, it also conserves water and is environmentally friendly. This product provides a convenient and safe alternative for homemakers as well as those who travel. Just use clean water with no detergent for cleaning infants undergarments and avoid the danger of chemical allergy and skin rash. This multifunction ultrasonic cleaning ball is very convenient for cleaning underwear[12], socks or stockings when traveling. It can also be used for delicate items that cannot withstand the churning of a washing machine. However, it is also very useful in the washing machine, for normal washes where it will not only add ultrasonic cleaning but will also stop washing from knotting up. Important clothing will not be damaged by the machine. In the past people often threw a few washing balls into the washing machine to prevent cloth from knotting up. This new ultrasonic washing ball has the added function of strengthening water flow. Just one ball can prevent clothing from knotting up and become hard to take out.

This device can also be used for cleaning in dangerous or narrow and small locations, such industrial chemical tanks or water towers where hand cleaning might be difficult or even hazardous. The washing ball makes cleaning more effective and improve labor safety, which can satisfy the requirements of disaster prevention and hazard mitigation.

References

[1] Riyas S and Das P N 2004 *British Ceramic Transitions* pp 23－28
[2] Birbrayer A 1998 *Seismic Analysis of Structure* (St. Petersburg: Nauka Press)
[3] Zlenko A A 2015 *Astron. Rep.* **59**(1) pp 72－87
[4] Akulenko L D, Markov Yu G and Rykhlova L V 2001 *Dokl. Phys.* **46**(4) pp 261－263
[5] Munk H, MacDonald G 1960 *The rotation of the Earth* (University Press) p 323
[6] Kozicki M and Sąsiadek E 2012 *Color. Technol.* 128 pp 251－260
[7] Akulenko L D, Markov Yu G, Rykhlova L V 2001 *Dokl. Phys.* **46**(4) pp 261－263
[8] Tyapin A 2013 *Analysis of Structures with Regard for Subgrade Interaction* (Moscow: ASV Press)
[9] Saunders M, Lewis P and Thornhill A 2009 *Research Methods for Business Students* (London: Financial Times Prentice Hall)
[10] Brettel, M, Friederichsen, N, Keller, M, and Rosenberg, M 2014 *International Journal of Mechanical, Industrial Science and Engineering* **8**(1) pp 37－44

[11] Uzdin A, Sandovich T. and Al'-Naser-Mokhomad Samikh Amin 1993 *Fundamentals of the Theory of Seismic Resistance and Earthquake-Proof Construction of Buildings and Structures* (St. Petersburg: Vedeneyev Press)
[12] McGuire, T W and Staelin, R 1983 *Marketing science* **2**(2) pp 161 – 191